INTRODUCTION

TO

SWITCHGEAR

HIGH VOLTAGE ENGINEERING

BY
HENRY E. POOLE

INTRODUCTION BY
GREG EASTER

Preface

To be fluent in the technology of any science requires knowledge of its historical development. To fully appreciate today's methods and materials, one must study the path which progress has taken. High-voltage switchgear has evolved over the decades to meet the needs for the vast increase in electrical power demand, concerns about safety and the environment, and prolonged low-maintenance operation.

Still, the basic theory and methods of operation have changed only in relatively small ways since the 1920's. After this introduction to some key differences in today's switchgear technology, there is a complete unabridged reprint of a rare early manual. The diagrams provided there are especially clear, and the underlying principles are identical to any system today.

Introduction

Switchgear in an electrical distribution system divides power into small circuits and provides protection against overloads by means of circuit breakers. These switches may be triggered to interrupt power automatically when an overload condition exists, as determined by thermal or magnetic sensors in an electronic trip unit, or under human control for whatever reason.

When excess current flows through a small element that heats it beyond a certain temperature, the breaker is tripped. Alternatively the strength of the electromagnetic field in a small coil may be used to detect abnormal conditions and trip the circuit by either electronic or mechanical means. In a modern industrial setting, the trip condition would be reported back to a monitoring display where the appropriate action would be taken. At some point this would mean resetting the circuit breaker. In earlier systems, this meant that a technician had to actually travel to the location of the breaker and manually reset it. Modern switchgear is more often reset by electromechanical means, such as a motor that can be activated from a control center. Solenoids are also commonly used. These electromechanical devices may also drive a piston that performs the reset by hydraulic pressure, depending on the design.

SCADA switches (Supervisory Control And Data Acquisition) are operated from a master station, just as circuit breakers may be reset without requiring anyone to be physically present. This improves both safety and efficiency. This topic is beyond the scope of this book. For an overview of this subject, point your browser to...

http://en.wikipedia.org/wiki/SCADA

The tremendous increase in demand for electrical power in metropolitan areas and modern industrial complexes has made it necessary to construct high voltage systems in the center of densely populated regions. Switchgear must be constructed to minimize electrical arcing for several reasons. These reasons include wear and tear on the switches themselves, the electrolytic production of toxic substances and gases (including ozone and nitrogen oxides from corona discharges in air) and finally, arcs generate electrical spikes that can trip other breakers in a cascade effect.

By 1920 the solution to these problems was the use of transformer oil, which was simply refined mineral oil. Because this is also flammable, overloads produced serious explosions with the oil quickly spreading the fire to the surrounding area. It was apparent that some type of nonflammable oil was needed. *Pyranol* and *Inertol* were trade names of two of the most common nonflammable oils used from about 1932 to 1978. These were both polychlorinated biphenyl (PCB) compounds.

As with mineral oil, PCB acts as both a dielectric and lubricant for the blades of the switch, which remain completely immersed in the oil. The replacement was not simply a matter of draining the mineral oil and replacing it with PCB, however. All transformers and switchgear had to be replaced with new designs that were resistant to the gummy residues that resulted over time, as well as the corrosive power of the chlorine gas liberated from the PCB in small amounts from high voltage electrolytic reactions. However, because PCB's are inexpensive, nonflammable and operate quite reliably with almost any type of mechanical switch or transformer, they became the unquestioned technology of choice the world over.

Unfortunately the PCB transformer oils that had been so popular were later discovered as the source of certain carcinogenic compounds in landfill sites, most notably dioxins and furans. Since PCB transformer oils need to be drained and replaced in most equipment of this kind, the amount of waste generated by their use was tremendous. The toxins eventually found their way into rivers and lakes, where the toll on wildlife, especially fish, continues to be devastating. By the 1970's there were tens of millions of transformers, switches and circuit breakers using PCB transformer oil. In 1977 the United States government banned the production of PCB transformer oils, forcing other materials to be substituted. However some of the original PCB equipment still remains in use, particularly in less developed nations where economics sometimes take precedence over public safety. Also in some older sealed equipment that has been in operation since before the ban.

There are other types of transformer oils in use, including silicone-based materials and fluorinated hydrocarbons that resemble liquid Teflon. Both of these are expensive when compared to the older PCB oils, although they appear to solve the issue of toxicity based on available data at this time. Even vegetable oils (e.g. canola oil) have been used with some success, but all of these approaches have drawbacks in cost, toxicity and/or performance.

More recent innovative mechanical designs have enabled switchgear to operate with air as the insulator instead of any oil, but this introduces some new problems. First, these air-insulated circuit breakers are very large, and a lot of them demand a great deal of space. This is often the deciding factor when the power distribution center is located in a city's center where real estate values are at a premium. Second, the air itself is frequently a problem in a modern urban environment. Air-insulated switchgear is sensitive to dust, pollution contaminants, moisture, salts, and even some gases. So another level of controls must be in place around these devices in order to filter the air, which adds to the expense both in the equipment itself as well as still more land and building space requirements. Finally, as previously mentioned, arcs in air produce ozone and nitrogen oxides, which are both highly toxic. So monitoring and other precautions must be taken when such equipment is to be located in the midst of a populated area, even with state-of-the-art technology that produces very low emissions.

There have been two popular solutions brought into practice. The first, and most common in recent designs is gas-insulated switchgear (GIS). At present this is almost synonymous with the use of sulfur hexafluoride gas, known as SF6 (from the chemical formula SF_6). The advantages include being nonflammable, of low-toxicity, and being quite inexpensive. In addition, even at very high voltages the effects of corona discharge and RF noise are completely eliminated. These SF6 units are generally equipped with an internal filter system that traps the traces of gas decomposition products in a holding chamber, thus reducing their maintenance. They are typically designed with one case inside of another case, and small bursting plates between these two chambers that will open in the event of excess pressure, thus containing the explosion within the outer barrier. This outer wall also prevents any person or animal from coming into contact with live high voltage conductors, which is especially important for equipment that will be situated in the middle of a city.

The other important new technology is vacuum-insulated circuit breakers (VCB). Their main advantage is that they are practically maintenance free. The disadvantages are higher cost and a 36-kilovolt maximum operating limit, at least at the present time. There are also "low-pressure" systems that employ SF6 gas at a partial

6

vacuum, being a hybrid between the two approaches to extend the voltage range. This is new and rather expensive technology. It will take time to see if there are unforeseen problems with this approach.

Both GIS and VCB switchgear offer a wider latitude of operating conditions than early oil-insulated designs could provide, and generally remain in continuous service for much longer with little or no maintenance. The switching time of these newer systems is in the range of a few cycles, thus improving the protection of equipment and reducing the chances of more serious damage from occurring.

It should also be mentioned that some switchgear is an integral part of other high-voltage equipment, and not as clearly categorized as simply bus switches and circuit breakers. An important example of this is the Buchholz relay, also known as the sudden-pressure relay. These have been in common use since about 1950, so they are not mentioned in the older reprinted manual that follows this introduction. Buchholz relays are found on liquid-insulated transformers and choke coils and serve to guard against a failure inside the device itself (as distinguished from merely a fault on the line). When a transformer overheats or it produces gas. A Buchholz relay will automatically respond to three types of failures. The first is slow gas accumulation, which occurs when there is a prolonged minor overload condition present. The relay adjusts the level of insulating liquid inside of the transformer or choke coil to displace the gas. The next type of fault that can be detected is a rapid flow of insulating liquid inside the device, which indicates high-voltage discharges that heat the oil and cause it to circulate by convection. The initial response of the Buchholz relay is to move dampers in position that restrict the flow. If there is too much pressure, then the circuit breaker is tripped. Finally, a loss of insulating fluid will also trip the breaker, whether in the event of a slow leak or a rupture. In any one of these instances a signal is sent to the main control station to alert the power plant operators of the fault condition. The relay contains a compartment with a sample of the gas that has been generated by the transformer, and a technician may withdraw samples for analysis. The following URL is a detailed description of a modern Buchholz relay...

http://www.emb-online.de/pdf/BR_ENG.pdf

All three systems, whether employing liquid oil, air/gas, or vacuum as the insulating medium, require a tripping mechanism to initiate the circuit break. There are three basic types, the spring-loaded (which may be reset by a motor drive for remote control), an electromagnetic solenoid, or a vacuum/pressure differential method. Simple classification is rarely possible today because by the many hybrid technologies, such as an electromagnetic solenoid which drives a piston to create a pressure differential, etc. Regardless of which mechanical principles are employed, the goal is to be able to break the connection as quickly and cleanly as possible. The connection is made (and broken) between two conducting surfaces, usually referred to as blades. Everything else that has changed since the 1920's is really just a refinement to the basic principle of dividing high-voltage transmission down to lower voltage circuits while guarding against fault conditions due to surges and equipment failures.

For more about the state-of-the-art in electromagnetic actuators for VCB switchgear, including the mathematical formulas for computing the delay in such systems, the reader is referred to Mitsubishi Electric ADVANCE Journal, volume number 116 (December, 2006) which may be viewed online at...

http://global.mitsubishielectric.com/company/r_and_d/advance/advance.html

The title of the article is, "Electromagnetically Actuated Vacuum Circuit Breaker."

So what else has changed? In the past the conductors were invariably copper and the casings were steel. Insulators were generally ceramic/porcelain or PCB oil-filled ceramics. Today casings are often aluminum or epoxy-coated steel. Conductors are also often aluminum, and insulators may be a composite fiber or epoxy resin material. This makes the entire unit lighter and provides for lower enclosure losses. Insulators are still ceramic/porcelain, and sometimes filled with sulfur hexafluoride gas in place of PCB transformer oils.

In the past the casings were equipped with small glass windows and other openings for visual inspection and access for the routine replacement of the insulating oil. Today's cast aluminum switchgear casings are often hermetically sealed to protect them from the atmosphere, and they remain in continuous operation for extended periods without the need for any maintenance. There are many types available today that are completely submersible and impervious to harsh environmental conditions.

Parts are usually modular in today's designs, and while they vary from manufacturer to manufacturer, the concept is the same in enabling different configurations to be built for a particular application at a minimum of cost. Since these parts can be disassembled for shipment, difficulties and expense in their transportation have been minimized. This also means that new equipment can be brought online more rapidly.

It should also be pointed out that a significant part of the world still uses oil-insulated switchgear that is very similar to the designs shown in the following reprinted book, "High Tension Switchgear."

For the seasoned electrical engineer, the following will provide a rare insight into the technical aspects of exactly how earlier types of switchgear operated.

HIGH TENSION SWITCHGEAR

DESCRIBING THE DESIGN, CONSTRUCTION, AND
FUNCTIONS OF THE LEADING TYPES OF SWITCH-
GEAR USED IN THE CONTROL OF HIGH-TENSION
ELECTRICAL PLANT

BY

HENRY E. POOLE
B.SC. (HONS.), LOND., A.C.G.I., A.M.I.E.E.

PREFACE

This volume is intended to give a general description of the leading types of apparatus employed in the control of high tension electrical plant. The more important points in the design are considered, but highly technical details are omitted. As far as possible controversial points have been avoided, and where these are included, both sides of the question are placed before the reader.

In most cases the actual apparatus described and illustrated are from the author's own experience of design, but care has been taken that the plant is either representative of a general type or illustrative of some matter of general interest.

HENRY E. POOLE.

London,
December, 1920.

CONTENTS

ILLUSTRATIONS

HIGH TENSION
SWITCHGEAR

CHAPTER I

INTRODUCTION

General Considerations. The great difficulty
in the design of high tension switchgear is the
variation in the conditions of operation.
Apparatus which is quite satisfactory for a
given current and pressure on a small system,
may be quite unsuited for use with a larger
plant.

An approximate idea of the severity of
operation can be obtained from an estimate of
the possible effect of a short circuit at the point
under consideration.

The current which will flow into a fault
obviously depends both on the capacity and
design of the apparatus connected to the
system, and on the resistance and inductance

1

existing between the point of breakdown and the source of supply.

For generating stations, the short circuit current may be estimated as approximately equal to three times the total normal full load current of the alternators connected to the bus bars. The momentary surge may be many times this value, but the liability to damage can be taken as roughly proportional to this figure.

If the apparatus is connected to the bus bars through a cable, transformer or other resistance or inductance resulting in a pressure drop equal to, say, n per cent. of the normal pressure, the short circuit current should be calculated by multiplying the full load current by the factor $100/n$. Thus, in the case of switchgear connected to the secondary side of a transformer having $3\frac{1}{3}$ per cent. drop, the short circuit current may be taken as thirty times the full load current unless this value is greater than the short circuit current of the generating station.

V.D.E. Rules. Up to the present the only regulations which have been published with regard to switchgear design are those prepared by the *Verband Deutscher Elektrotechniker,* or German Institute of Electrical Engineers*. It has been stated that these rules are merely a summary of continental practice, and do not rest on a scientific basis, but the V.D.E. declare that they are the result of actual tests. Other authorities, such as the American Institute of Electrical Engineers, content themselves with general statements without giving actual figures, the only definite data other than those contained in the German rules lacking any official backing.

It may be taken that any gear designed to V.D.E. Rules is quite satisfactory, as the values specified are greatly in excess of those employed in current British and American practice. Continental firms frequently specify two working voltages, one in accordance with the standard ratings, and the other a " highest permissible working pressure," which is often as much as double the more conservative rating.

The V.D.E. Rules specify three *gap dimensions,* of which two only apply to oil switches.

Dimension *A* is the least distance in air

---

* The British Engineering Standards Association have a specification of oil switches in preparation. The principles adopted are very similar to those of the V.D.E. Rules.

between bare current-carrying parts and earth, or between phases, or if the apparatus is switched off, between the various parts of the same pole of phase.

Dimension B is the least clearance under oil to earth, to the surface of the oil, between phases, or when the apparatus is switched off, between the various parts of the same pole or phase. This dimension applies only to oil switches, and not to such parts of the switch as are outside the influence of the arc.

Dimension C specifies the minimum depth under the surface of the oil to which the breaking point of an oil switch should be immersed.

The figures given are as shown in Table I.

TABLE I.—MINIMUM GAP AND SUBMERSION DIMENSIONS (V.D.E. RULES).

Normal Pressure volts.	Short-circuit Current. Amp.	Dimension A In.	Dimension B In.	Dimension C In.
1,500	Up to 6,000 }	3	$1\frac{9}{16}$	$3\frac{1}{2}$
3,000	Up to 3,000			
3,000	3,000 to 6,000 }	$3\frac{15}{16}$	$1\frac{31}{32}$	$3\frac{15}{16}$
6,000	Up to 3,000			
6,000	3,000 to 6,000 }	$4\frac{15}{16}$	$2\frac{3}{8}$	$4\frac{3}{4}$
12,000	Up to 1,500			
12,000	1,500 to 4,500 }	$7\frac{1}{16}$	$3\frac{1}{2}$	$7\frac{1}{16}$
24,000	Up to 1,000			
24,000	1,000 to 2,000 }	$9\frac{7}{16}$	$4\frac{3}{4}$	$9\frac{7}{16}$
35,000	Up to 1,000			

The short circuit currents are calculated by the methods outlined above.

Apart from the large dimensions, the V.D.E. Rules depart from current British practice in one or two particulars, notably in allowing the same dimension to hold for sparking distance through air and creepage over insulation. It must be remembered, however, that any such regulations as these must err on the conservative side so as to ensure safety if they are to be followed blindly regardless of individual circumstances.

Testing pressures, and various details of design, are also mentioned in the V.D.E. Rules, and will be considered in due course.

CHAPTER II

OIL SWITCHES

General Considerations. The performance of an oil switch depends on the carrying capacity of the contacts and conductors ; on the clearances to earth between phases both in air and in oil ; on the length and speed of break ; on the depth of immersion of the breaking point under oil and the air space above the oil ; and on the capacity and mechanical strength of the oil container.

The switch should be designed with due regard to balance between these various considerations. It is very easy to exaggerate certain features without any benefit to the apparatus as a whole. For instance, on one system it was found that the breaking capacity in the oil tank was quite satisfactory, but that on short circuit the switch flashed across the terminals.

The influences of the *carrying capacity* and *clearances* are obvious. With regard to the latter it must be remembered that owing to pressure surges due to switching, resonance effects, lightning discharge, or other cause, a large factor of safety is required above the normal working voltage.

The *length of break* determines to a great extent the rupturing capacity of the oil switch, but it should not be increased to such an extent that current tends to jump across other parts, such as between the terminals. For a double break switch, if the distance between the nearest point of the moving element and the fixed contacts is equal to the other clearances under oil in the vicinity of the arc, and the minimum sparking distance in air is about double that value, the design may be considered to be correctly proportioned.

Some difference of opinion exists as to the influence of the *speed of break* on the performance of an oil switch, and it is to be noted that no figures for this are specified in the V.D.E. Rules. Some engineers are of opinion that since the arc must in any case persist for several cycles the time taken to break circuit can be varied between large limits without adverse effect, while other engineers consider the matter to be of greater importance. In any case the time interval is not of such moment with an oil switch as it is with an air-break circuit breaker.

When a current is broken under oil, bubbles of gas are formed by the arc, and the arc itself is extinguished by the oil rushing between the contacts. The influence of the *depth of immersion* of the breaking point is to increase the oil pressure on the break, and so diminish the volume of gas formed, and to give a quicker flow of oil to extinguish the arc. Another point is that the gas formed is inflammable, and if no oil intervenes between the arc and the atmosphere, there is a danger of ignition taking place. The great importance of the head of oil above the breaking point is thus apparent.

A certain volume of *air above the oil* is necessary on account of the explosive nature of the rupture of the current. The air forms a cushion to absorb the shock, and prevents mechanical damage resulting. In any case, when the switch breaks under short circuit conditions, considerable stress is set up in the oil container.

The oil takes the shock in the first instance, and so the *volume of the tank* influences the breaking capacity of the apparatus. However, in some designs the oil container is purposely kept small, so as to limit the after effects of an explosion.

Types of Switches. Oil switches are made

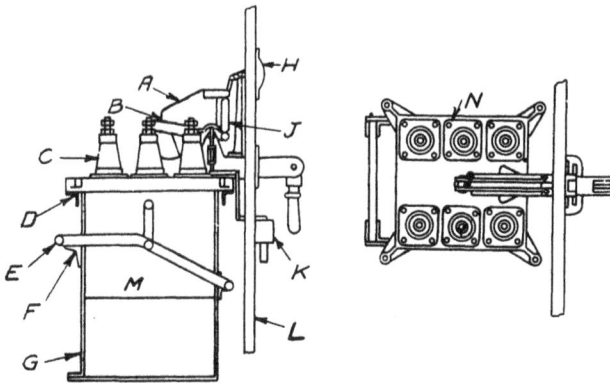

FIG. 1.—HIGH-TENSION OIL SWITCH.

A. Trip case.
B. Switch main lever.
C. Thoroughfare insulator.
D. Switchboard frame angle
    supporting switches.
E. Tank lowering lever.
F. Catch holding tank in position.
G. Tank guide frame.
H. Switch position indicator.
J. Spring.
K. Trip coil.
L. Panel of switchboard.
M. Tank.
N. Detachable insulator flange.

in single, double, triple, and four-pole forms, of which the triple-pole form is most in use. For cellular-type switchboards, where the phases are kept entirely separate throughout the board, three independent single-pole units are coupled together mechanically, to form a triple-pole unit (Fig. 5).

Switches for very high pressures sometimes have more than two breaks per pole, but even for extreme voltages, it has been found that more than four breaks per phase are not advantageous.

The most usual design of oil switch has the stems brought through thoroughfare porcelain insulators mounted in the top of the switch. Contact is made by a moving element immersed in an oil tank (Figs. 1, 2, 5, 7, 8 and 12). Another design has thoroughfare insulators in the bottom of the tank, the moving element moving upwards to break circuit (Figs. 3 and 4).

In the more usual type the switch top is of cast-iron for moderate currents. For currents above 1,000 amp. brass is used to a great extent for high tension switches, and slate tops without porcelain insulators for lower pressures. Pressed steel tops are sometimes employed for large switches, where a cast-iron top is inconveniently bulky. This type gives some of the electrical advantages of a brass top, as although steel is used the bulk of metal is small, and so the magnetic and eddy current losses are kept within limits.

Thoroughfare Insulators. In the early days, thoroughfare insulators were almost always corrugated. The grooves greatly increase the

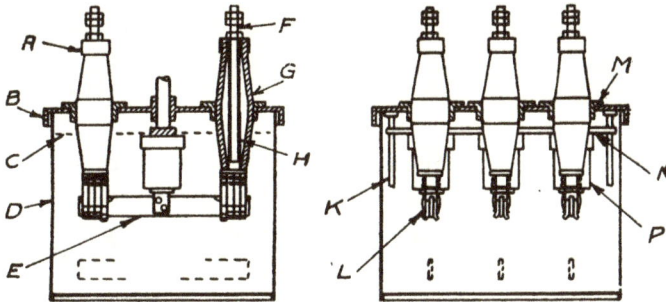

FIG. 2.—OIL SWITCH INTERIOR.

A. Insulator cap.	H. Mica sleeve.
B. C.I. switch top.	K. M.S. guide rod.
C. Oil level.	L. Contact fingers.
D. M.S. tank.	M. Detachable fingers.
E. Copper blade.	N. Moving element bridge.
F. Copper switch stems.	P. Pot insulator.
G. Hollow insulator.	

creeping surface for a given height of porcelain, and so increase the breakdown pressure in a laboratory or workshop test. However, under practical conditions, the upper horizontal portions of the grooves become covered with dust, thus leaving the duty of insulation to the other parts. Under these conditions, the actual insulating value of the porcelain becomes less than that of a plain insulator.

The thoroughfare insulators are fixed either directly in the switch top, or in detachable flanges bolted to the switch. Litharge cement and low melting point metal are the fixing mediums employed. Another design has a plate clamping a flange on the insulator, a washer of soft material distributing the pressure.

Litharge cement has the advantage that, if the proportions of litharge and glycerine are correct, the mixture expands very slightly on setting, and so holds firmly. The quality of the materials used is of great importance, both with regard to the first setting and with reference to the possibility of deterioration with age. Type metal makes a very sound mechanical job, but may crack the insulator unless poured at the correct temperature, with the porcelain preheated and free from moisture.

One of the difficulties with insulators is that they cannot be made with the same accuracy of dimensions as other materials used in engineering. As regards diameter the defect

can be made good by varying the thickness
of cement, but variations in length cause
more trouble.

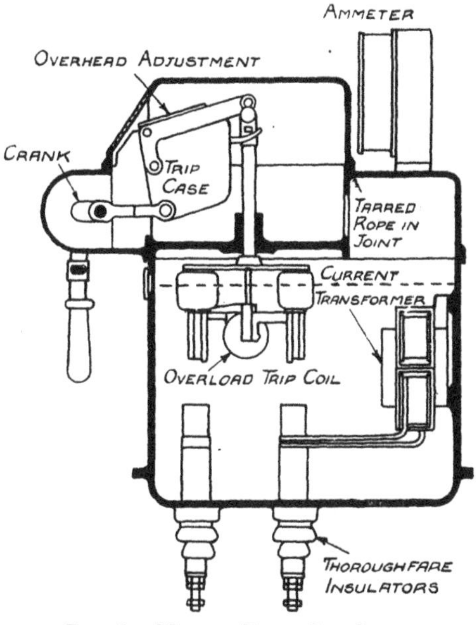

FIG. 3.—MINING TYPE OIL SWITCH.

An easy way of getting over the lack of
accuracy is to make the insulator without
flanges, and cement the porcelain to jig with
the bottom at a certain distance from the
switch base. The variation in the projection

of the stem causes but little inconvenience, (Figs. 1, 2, 5, 7, and 12).

At first sight this arrangement appears to place too much reliance on the strength of the cementing medium, but the stress of switching-in is taken by the fixing even with a flange in the conventional position above the switch base, and so there is little disadvantage in letting the approximately equal switching-out stress also come on the cement. The weight of the insulator and stem is, of course, small in comparison with the switching stress.

The distance between the live stem and the earthed metal of the switch base is increased by enlarging the diameter at the centre of the thoroughfare insulator, but porcelain cannot be guaranteed free from flaws when above a certain thickness. For this reason, hollow barrel-shaped insulators possess many advantages, especially for the higher pressures. The switch stem itself is sometimes additionally insulated by a mica sleeve, or the whole interior of the insulator may be filled with oil or joint-box compound, as these materials have a higher dielectric strength than air.

Switch Stems. For currents up to 1,000 amp.

the switch stems are almost always circular in section, the connections being clamped between nuts on the stem above the switch. In one design, however, the cable thimble is brought right inside the insulator.

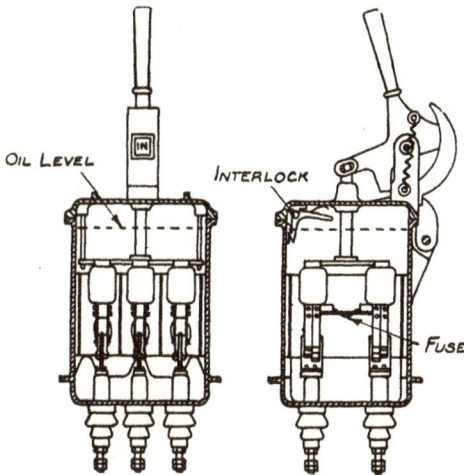

FIG. 4.—OIL SWITCH FUSE.

The stems are frequently screwed with a finer thread than the standard Whitworth, to increase the carrying capacity by obtaining the greatest possible section of metal under the thread. The nuts are often larger in size than standard to give a good clamping surface, but they can be made thin as the mechanical stress in the thread need not be excessive. For instance, a ½ in. diameter stem may be threaded ½ in. B.S.F., and fitted with a ⅝ in. standard half-thickness nut tapped to suit.

For small currents brass stems are employed instead of copper, as for mechanical reasons the dimensions should not be brought below a certain minimum. Stems ⅜ in. diameter are quite small enough for any high tension switch, and for high pressures ½ in. is about the average low limit.

Circular stems are also used for large currents in high tension switches, as they can conveniently be brought through porcelain insulators. In one design, in order to get a flat clamping surface, the current density is kept low through the switch top, and above the insulator half the stem is cut away. The connection is clamped to the vertical surface thus obtained by bolts passing across the stem, or by studs tapped into the copper. In another arrangement a flat strip is sweated into a slot cut in the centre of the stem to give a convenient connection.

The difficulty with large circular stems is chiefly that of connection to the other gear in the switch-panel. The lower end of the copper rod can be sweated into a brass casting carrying the contacts, this arrangement offering little difficulty as regards carrying the current.

When the pressure permits of the employ- ment of a slate switch top, flat copper strips make a very convenient connection to the contacts. These strips are generally insulated from the slate by presspahn, or moulded mica wrappings, so as to minimize the effect of metallic veins in the slate.

For normal currents the carrying capacity of the stem is usually calculated at the con- ventional 1,000 amp. per sq. in., but when the current exceeds about 1,500 amp. the current density has to be reduced, the generally accepted figures falling in proportion to the current, to about 600 amp. per sq. in. at 4,000 amp. normal load. Heavy-current switch stems are frequently laminated to break the path of eddy currents.

Stems should be fixed rigidly at one end only when mounted in insulators, so that the difference in the coefficient of expansion of copper and porcelain may not cause any mechanical damage if heating takes place. The end fixed is generally that near the contacts. The upper part of the stem can be brought through a metal cap fitting the top of the insulator, and held by a nut pressing on the cap, other nuts clamping the connecting strip or thimble. The lower end can be tapped into a brass casting cemented into the base of the porcelain and carrying the switch contacts.

Contacts. The most simple form of switch contact consists of copper fingers backed by steel springs, pressing on a flat copper switch blade (Figs. 2, 5, and 12). As in the case of switch stems, brass is employed for mechanical reasons, where the higher current carrying capacity of copper is not required. These fingers are usually rated at 75 amp. per sq. in., on the contact surface. Controller type contacts are also employed.

A finger which is much in use is $\frac{1}{2}$ in. wide, the contact being $\frac{1}{16}$ in. thick, riveted to a $\frac{1}{2}$ in. by $\frac{1}{16}$ in. copper blade. The clamping surface on the finger-block, the distance from the block to the top of the contact and the length of the contact itself, are each 1 in. The end of the contact is extended $\frac{3}{8}$ in. and bent through 45 degrees to allow the blade to enter

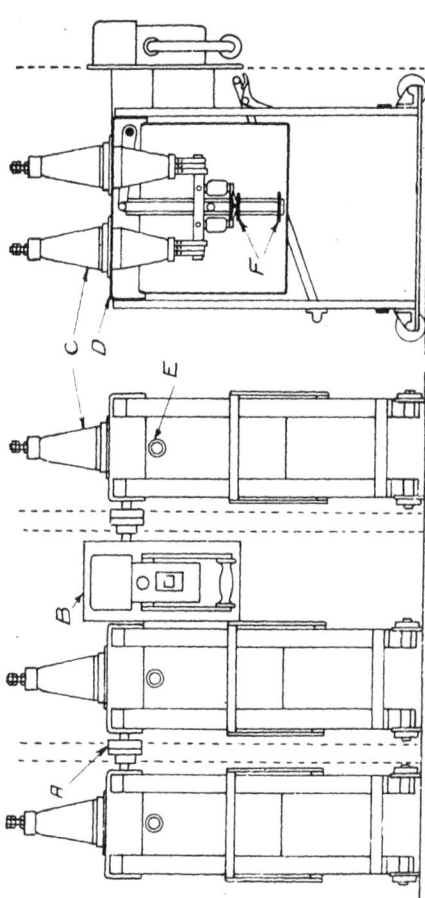

FIG. 5.—Extra High-tension Oil Switch. Three Single-pole Switches Coupled Mechanically.

A. Dismountable coupling.
B. Electrical operating gear.
C. Porcelain insulators.

D. C.I. switch top.
E. Oil level inspection glass.
F. Motion retarding plates.

easily. A backing spring $\frac{1}{2}$ in. by $\frac{1}{32}$ in. in section is employed, two $\frac{3}{16}$ in. screws with lock washers clamping both finger and spring to the contact block.

In order to ensure contact over the whole surface, and so increase the carrying capacity up to 100 amp. per sq. in. the fingers are sometimes made self-aligning, though this introduces some mechanical weakness.

In self-aligning fingers the contact surface and the backing plate are made separate, the spring pressing on a button in the centre of the contact, which passes through a clearance hole in the backing plate, with another pin to help alignment. The current is conveyed to the contact through a flexible connection, either of stencil copper or dynamo brush flex.

The plain blade design is quite satisfactory up to about 600 amp., but above this current the friction on the blade is excessive and makes it difficult to arrange for a convenient tripping arrangement.

For currents above 600 amp., and in many designs for values below this figure, the blade is made an inverted V-shape, the slope of the sides being about 15 degrees from the vertical (Fig. 6). The blade may be made from two

flat copper strips bolted to a supporting casting, or may be of stamped metal. Cast
copper blades are also employed, but there
is sometimes difficulty in getting this material
free from defects.

The current is conveyed to the blades by
self-aligning contact fingers. This design has
the great advantage that the pressure of the
fingers tends to trip the switch. The contact
obtained is so excellent that the current
density may safely be increased to 125 amp.
per sq. in., and by the use of several blades
in parallel, any practicable current can be
dealt with.

Brush type contacts are also employed,
especially for the larger currents. The current
may be carried either by an inverted U-
shaped brush, reinforced by backing springs,
and bridging flat horizontal surfaces at the
base of the switch stems, or by separate brushes
carried on a connecting bar. The allowable
current density varies greatly with the design,
values from 60 to 250 amp. per sq. in. being
employed.

Another type of contact is provided by a
vertical brush bearing on the sides of a V-
shaped blade, but this type requires very

accurate adjustment of the height of the moving element if contact trouble is to be avoided.

As compared with finger contacts, brushes give excellent current carrying capacity when in good condition, but they are very easily upset by lack of alignment or mechanical defects generally, and give trouble if the surface is burnt in any way in breaking circuit.

Brush contacts have the advantage of giving a large break for a given motion of the moving element, but they have the disadvantage that the break tends to take place when the switch is moving slowly at the commencement of the switching operation, whereas finger contacts rupture the current when the speed of movement is at its maximum and so the break is more effective.

As a matter of fact, the smallest sparking distance under oil in an oil switch of moderate size is generally that between the contacts and the cross-bar of the moving element, and there is no great practical advantage in making the break more than this, though a long break is often a good " selling point."

Sparking Gear. In all switches, except

those for very small currents, an attempt is made to open the circuit away from the main contact surfaces. For a single blade this is easily secured by making one of the fingers $\frac{1}{4}$ in. or more longer than the others, but this is only satisfactory for moderate currents (Figs. 2, 5, and 11).

For currents over about 600 amp. and for severe conditions of operation, such as those to which rolling mill switches are subject, other forms of sparking gear are necessary.

Separate sliding contacts are often provided at the end of the blade to take the sparking, but these suffer from the defect that as soon as a bead of metal is formed by the arc, this projection forms the only contact surface.

The best form of sparking gear is provided by two substantial blocks of metal butted together (Fig. 6). These blocks remain in contact at the commencement of the switch motion, but are parted when the moving element is at about its maximum velocity. The speed of break, and the comparatively large bulk of metal, both tend to reduce the effect of the arc, and give excellent results. For a multi-blade switch several of these contacts are provided.

The current must be conveyed to the moving sparking contact by a flexible connection to prevent any danger of arcing welding the moving parts together, and also to ensure that the butt contact takes the current when the main circuit is broken.

Oil switches have been made with the butt type of contact only, but the current carrying properties of such switches are not good, because, if a bead of metal be raised by the arc when the circuit is broken, the top of the bead is subsequently the only point in contact. Flat copper blocks are theoretically the ideal contact surfaces, but they are much too easily upset to be of any practical use.

For crushed brush switches an additional leaf, tipped with copper or non-arcing metal, is often arranged to break contact after the main switch.

Magnetic blow-outs or carbon breaks are not used on oil switches, the first on account of the switch almost always being used for alternating current, and the latter on account of the danger of carbonizing the oil.

Blade Supports. Single switch blades are generally attached to a casting having a boss

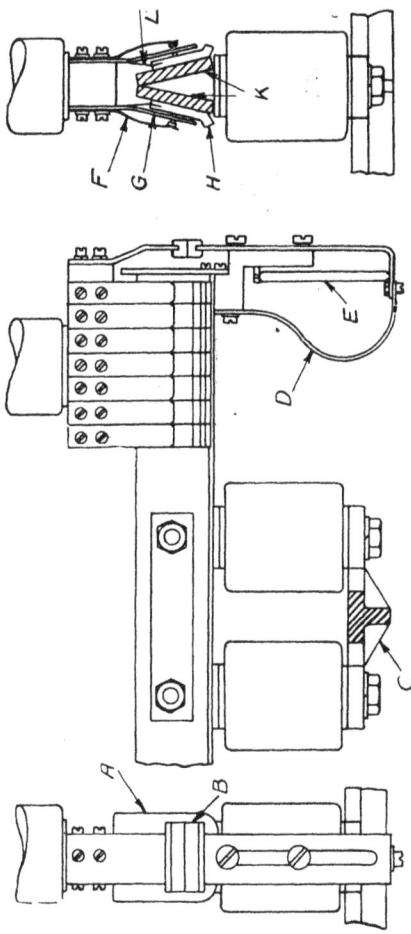

FIG. 6.—OIL SWITCH CONTACT.

A. Asbestos arcing shield.
B. Butt contacts.
C. Moving element casting.
D. Flexible connection.
E. Spring.
F. Finger spring.
G. Finger backing plate.
H. Finger contact block.
K. Copper switch blades.
L. Flexible connection.

for cementing into a porcelain insulator (Figs. 1 and 8). The insulator may be connected to a boss on the cross-bar of the moving element, or fixed inside a cup-shaped casting surrounding the base of the porcelain. To permit replacement to be made without the use of cement, the insulator is sometimes made with bosses cemented in each end, the attachment to the blade support and moving element being by set screws tapped into the bosses.

The insulators are assembled to jig with the faces of the metal at a standard distance apart, so that the alignment of the switch is not affected by replacement.

Pressed copper V-blades are generally clamped round a cone shaped projection on the insulator itself, to avoid having to trust to cement.

The pressure of both butt and V-shaped contacts tends to trip the switch, and so the operating force is always in one direction, both for opening and closing the contacts. When porcelain insulation is employed the insulators are generally placed below the moving element (Fig. 6) so as to be in compression, since porcelain is not very satisfactory in tension. For single blades nothing is gained by this construction, as the forces employed in switching in and switching out are approximately equal.

Two insulators as far apart as possible are sometimes used to prevent cross strain in the porcelain, and from considerations of electrical clearance these are below the blade (Fig. 5).

A moving element bridge-piece below the blade has considerable electrical advantage, as there is no earthed metal between the contacts under oil, but the depth of the tank and the difficulty of operation are both increased. The moving element bridge-piece above the blade can be used in many instances as the clearance necessary in air is about double that required under oil, and apart from this the earthed metal of the operating gear is frequently between the poles above the switch, and so keeps the contacts apart.

Moving Element Operation. In some cases blades are operated by separate impregnated wooden rods, and in another design the moving element bridge has moulded mica insulation. Other materials, such as paxolin, are also employed, but porcelain is, generally speaking, most in use for high tension work, especially for the higher pressures.

If wood or paper in any form be used, it should be treated so that it will retain its insulating properties. Unless special precautions are taken, these materials absorb moisture and so lose their dielectric strength after a time.

Moulded mica should not have too much varnish embodied in it or there is danger of softening with heat. The well known British Admiralty specification calls for a very high quality of this insulating material.

In order to secure better alignment, brush contacts are frequently allowed a little rotary movement on the moving element, to permit them to take up a position giving best electrical contact.

In small switches the moving element is attached to a plunger rod passing through the top of the switch, and this arrangement is applicable to single-pole switches, even in the larger sizes (Figs. 2 and 8). For big triple-pole assemblies two levers fixed to a rotating shaft at the side of the switch give much better results. The levers operate the

moving element through links at either
end of the tank. This arrangement is also
employed for single-pole assemblies (Fig. 5).

When used, the operating plunger forms the
principal guide for the moving element, supple-
mentary rod guides preventing rotation. With

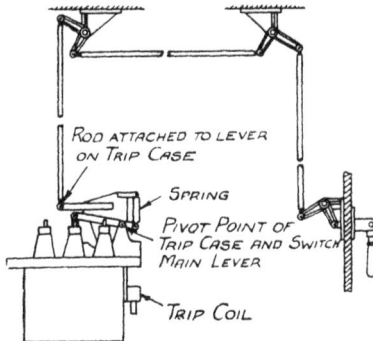

FIG. 7.—ROD OPERATED OIL SWITCH.

the lever and link motion the guides have a
much more important duty to perform, and
they are either in channel form with the ends
of the moving element forming a slipper block
inside them, or they consist of rigid and strongly
supported rods.

Cast iron can be used for the moving element
of small current switches, but material better
able to resist strain and shock must be employed

for heavier currents. Generally speaking the mechanical design of high pressure switches is simple, as the forces involved are small in comparison with the bulk required to give the necessary electrical clearances. The design of heavy current switches on the other hand is mainly a mechanical problem, as the surfaces to be kept in contact are relatively large, and the forces to be dealt with are considerable.

Oil Tanks. Oil tanks are generally made from mild steel of gauge varying from 16 S.W.G. on the smaller sizes to 12 S.W.G. on the larger. Of late years there has been a tendency to increase the thickness of the oil-container to enable it to withstand the stresses set up on short circuit. Small sized tanks are now frequently made $\frac{1}{8}$ in. thick, and boiler plates are used for switches for more exacting duties.

Switch tanks are generally fitted with lugs for bolting to the base casting. Except when small switches are in question, some form of tank-lowering gear is a very great advantage, and becomes absolutely necessary when the oil content is over, say, 10 gallons.

Tanks may be lowered by detachable levers hooking into projections on the tank and switch base, or by levers forming part of the switch itself. In the latter case the levers may be employed to hold the tank in position under load, the ordinary fixing bolts being dispensed with (Figs. 1 and 5).

Wire rope operated by worm gear is employed when the weight is too much to be handled conveniently by levers. This arrangement has the advantage that, owing to its irreversibility, it is not possible to drop the tank accidentally. When wire rope is used the tank is supported, when under load, by the usual bolts or by supporting levers.

In the case of very large switches the rope can be brought round under the tank, running on rollers at the bottom corners, as this arrangement halves the stress on the rope. The more usual sizes can easily be supported by two wires. Some form of adjustment must be made to equalize the load on the ropes.

Boiler plate tanks are frequently bolted to the top of the switch by angle-irons welded to the sheet metal, and the top of the switch is made strong enough to stand considerable stress.

Mining switches either have cast iron tanks or tanks made of boiler plate so as to stand rough usage, even when the conditions of operation are not otherwise onerous (Fig. 8).

Some designers use separate tanks for each phase in all switches with the object of preventing breakdown between phases under oil, while others prefer to use single tanks except for the largest switches, where the bulk becomes too great to be handled conveniently. The argument in favour of the single tank is that there is only insulating material between the phases, while the separate tanks increase the liability to break down for a given size of switch by placing the earthed sides of the oil containers between the terminals. For cell work the single pole units are, of course, entirely independent from the electrical standpoint.

The V.D.E. specify separate tanks for 24,000 v. when the short circuit current is over 1,000 amp. or for pressures over 35,000 v. whatever the working conditions.

Tanks are frequently lined with impregnated wood. Oil forms the best dielectric, but it is liable to be displaced by the vapour formed in breaking circuit. It is an open question whether it is better to increase the dimension to the side of the tank to a figure which prevents the risk of earthing, or to insert a lining to prevent a fault developing. If an explosion occurs, the smaller the tank the less burning oil there is to be thrown about, but on the other hand the oil absorbs the shock of breaking circuit in the first instance, and so the greater the cubic capacity of the tank, the less the liability to trouble. Switches for very high pressures are always lined, as otherwise the bulk becomes excessive.

The temperature of the oil under working conditions is of great importance, as there must not be any risk of ignition occurring. The V.D.E. specify that for switches up to 350 amp. rated capacity the rise is not to exceed 36° Fahr. ; for 2,000 amp. 54° Fahr. is allowed ; and for over 2,000 amp. the value given is 72° Fahr.

The quality of the oil used greatly influences the operation of switches. Ordinary good quality transformer oil, with a moderately high flash point and free from moisture, gives good results where the conditions are not too severe. However, the breaking capacity can be increased considerably by using a special quality white oil with a very high flash point prepared especially for use in oil switches.

Some means of indicating the level of oil in the tank should be provided. When the switch is operated the oil tends to carbonise and form sludge in the bottom of the tank. Arrangements must be made so that the sludge may be run off, or the oil completely changed, particularly if the switch operates on a direct current circuit.

The special switch oil has the advantage of being free from sludge, and so need not be changed so frequently as ordinary qualities.

Oil switches are almost invariably used on alternating current circuits, but they work satisfactorily with continuous current. The carbonization of the oil is considerably greater in the latter case, hence plenty of space must be provided for sludge to collect at the bottom of the tank.

Some means of escape for the gases generated in switching are essential. For central station switches of moderate size this can be provided quite conveniently by a simple cast iron hinged flap, which also forms a means of filling the tank with oil. The switch tank does not usually fit very closely to the top casting, and this provides additional paths for the escape of gas.

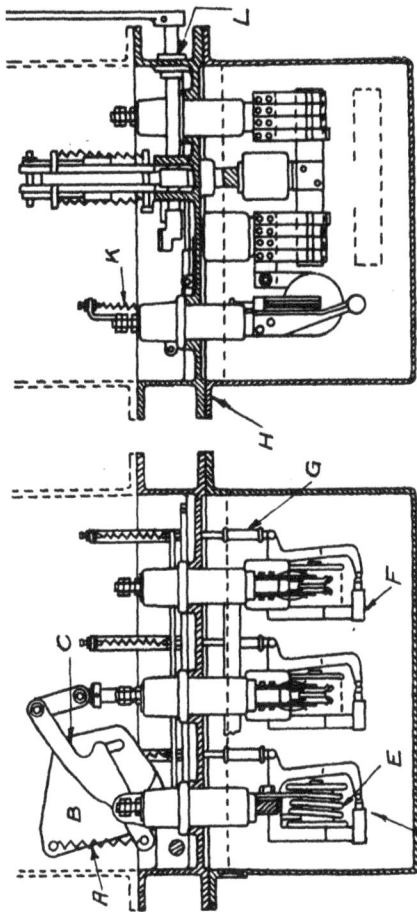

FIG. 8.—MINING TYPE OIL SWITCH WITH SUBMERGED OVERLOAD RELEASES.

A.   Switch main spring.
B.   Trip case.
C.   Switch main lever.
D.   Contacts omitted to show trip coil.
E.   Overload winding.
F.   Oil dashpot time lag.
G.   Trip rod insulator.
H.   Flameproof joint.
K.   Trip adjusting spring.
L.   Operating shaft bush.

Mining switches must be flameproof. One method of preventing the ignition of gas is to pack the joints between the tank and the rest of the switch and provide a flame cap with fine copper gauze similar to that used in a Davy lamp (Fig. 3).

The flame cap usually has some form of automatic valve which can be opened by the gases from the inside, but which prevents dust or other injurious material from entering the switch.

Another type of mining switch has wide machined joints bolted together (Fig. 8). The flanges should be machined accurately, but the tool marks left, so that the gases can escape between the surfaces. The cooling effect of the mass of metal insures that the gases leave at a temperature below the ignition point of the contaminated air of the mine.

For very severe conditions of operation, when the switches are provided with boiler plate tanks, the inflammable material is led right away from the switchboard through gas barrel, and discharged where no damage can be caused.

Charging Contacts. When a high tension

current is switched on, considerable electrical shock is brought on the insulation of apparatus on the dead side. In the case of transformers or induction motors the winding may be at earth potential, and the pressure in the first instance comes on the ends only.* The whole voltage of the line is thus brought across the first few turns, and although the risk of breakdown can be reduced by putting additional insulation on these turns, a better remedy is to switch on more gradually. With transformers there is also a considerable momentary rush of current under certain circumstances owing to the lag in the induction effect before the flux builds up, and cables and in fact all types of apparatus are strained by sudden switching.

The shock effect is greatly reduced by fitting the oil switch with potential charging contacts,

---

* When the current is switched at the peak of a wave, the pressure is applied to the transformer suddenly instead of gradually rising in sine wave form. For a fraction of a cycle no current can flow through even the very small impedance of one turn of the transformer winding, and for this period the full pressure is on the end of the winding while the next turn is still at earth potential. A more sudden rise of pressure than that given by the normal sine wave is equivalent to the imposition of a voltage of higher frequency than the normal supply, with corresponding phenomena as regards impedance effect.

by means of which contact is made in the first instance through a high non-inductive resistance.

A convenient resistance medium for potential charging consists of an asbestos strip with high resistance wire wound from side to side. This gives a great length of wire on a comparatively short length of strip.

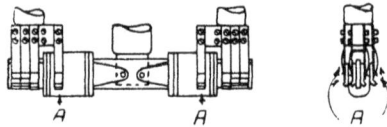

FIG. 9.—POTENTIAL-CHARGING CONTACTS.

In one form of potential charging contacts for single blades a sheet of copper or brass is wrapped round the blade, a lug passing under one of the blade supporting bolts. The end of the resistance strip is soldered to this, and the strip with a sheet of empire cloth on either side is wound round the blade. The other end of the strip is soldered to a wrapper of sheet copper which is bound round with wire sweated on. Contact is made by two additional fingers A, A (Fig. 9) packed out from the main finger block, and making circuit before the other fingers.

When there is not space on the blade, or when other types of contact are employed, the strip may be wound on tube, and disposed of elsewhere in the switch. Fingers insulated from the main block by mica plates and bushes make the required first connection.

In other designs separate stems are provided, the resistance being external to the switch, and yet another type has the resistance strip placed in the bottom of the tank.

The buffer resistance need not have any great heat radiating capacity, as the material is only in circuit for a fraction of a second during the process of switching, but the momentary rush of current is considerable.

Opinions vary greatly with regard to the ohmic value desirable for potential charging gear. A resistance of 1,000 ohms per phase, divided into two coils, one on either side of the blade, with resistance formed of 0.2 mm. resistance wire, gives good results with a 6,000 v. switch. For other pressures proportional resistances and sections may be used.

Switch Operation. Oil switches may be operated by a simple lever handle coming through the front of the switchboard. The lever is held in the closed position by a catch in front of the board, and tends to move into the off position under the action of springs and the weight of the moving parts of the switch. In the case of an automatic switch the catch is put out of action by the tripping mechanism.

With this arrangement, if the switch is tripped while the handle is held by the operator the motion is unaffected. In order to give a free-handle device, so that the switchboard attendant cannot hold the switch closed when it should be opened by the tripping gear, the operating handle is arranged to work the switch by an auxiliary lever through some form of catch. The catch is operated by the tripping gear independently of the operator.

This gear is simple for small units, but for large switches the force required to close the switch becomes excessive unless the arrangement is complicated by additional rollers or levers.

Switches are also operated by handwheels. These are very convenient for change-over switches, but for the more usual type they have the defect that the position of the operating gear is not very apparent, although a pointer is always provided.

An excellent type of switch operation is given by a handle fixed in a bracket projecting through the panel. The operating lever moves through 180 degrees, being vertical in the on and off positions. The switch is moved by a link passing through the board, a simple bell crank giving the requisite vertical motion in the case of a non-automatic switch. When tripping gear is required the link operates a trip case, and this picks up the lever which actually moves the switch (Fig. 1).

The power given by trip mechanisms is not very great, unless an excessive amount of energy is used to operate them. On the other hand it is an advantage to have springs of considerable strength to give a quick-break motion to the switch. Some form of power reducing mechanism is thus required.

A simple form of trip gear is provided by three levers. The first lever is in the form of a trip hook, with part extended to give a force ratio of about 4 : 1. The end of this lever engages in a second member which again reduces the force, and the end of this second lever is supported by a third. By careful design the force required to trip a switch may be reduced to a few ounces, even when the larger sizes are in question (Fig. 10).

Each of the levers is under the influence of a small coiled spring inside the trip case, which tends to bring it into the " switch tripped " position independently of the action of the moving element. The main switch springs can conveniently be arranged between a point on the trip case and the switch main operating levers.

When the switch is tripped the operating handle and trip case remain in the " in " position. To engage the mechanism the handle is moved into the " off " position, which brings the trip case up against the main switch lever, lost motion being taken up by a spring which also forms a buffer to retard the moving element. The first lever is forced into the " in " position, and pushes the second lever into engagement with the third. On moving the main switch handle into the " in " position the switch is again closed. With this mechanism a projection must be left on the second lever to prevent it from engaging with the third unless the first is also held, otherwise the first lever cannot push the second out of the way to engage itself, and the switch cannot be operated. This trouble can only occur if the switch is treated so roughly that the levers bounce into engagement, but when it does happen the trip case has to be taken to pieces.

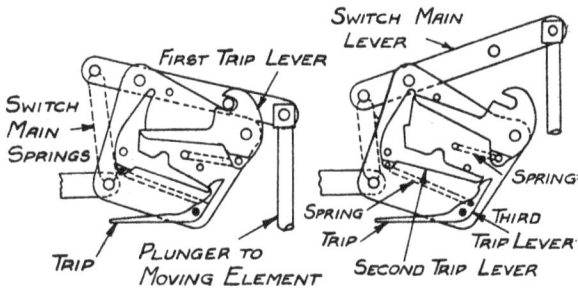

FIG. 10.—THREE-LEVER TRIP GEAR.

However, the simple addition of the projection effectively prevents any difficulty arising. This type of trip case can be applied both when the trip gear is behind or in front of the board.

Some form of quick-break mechanism is a very considerable advantage. This may take the form of a detent spring or lever attached to some fixed point and placed in the path of a tripping lever so as to trip the switch when the mechanism first commences to move. Some means of moving the detent out of the way when the switch mechanism continues its motion must be adopted to prevent breakage of the lever, and the trip lever must pass the detent without tripping when the switch is moved towards the closed position.

A more simple form, which has a limited application, is arranged to push the moving element out of the closed position by means of a spring, the motion being prevented at first by some form of friction catch. When the spring has been stressed to a certain extent the friction catch releases, and the motion parts operate rapidly under the influence of the stored energy.

In some trip mechanisms the tripping can only take place when the closing operation is completed. If the switch is used correctly this is not a very great disadvantage, but if the closing of the switch is performed slowly trouble may be caused if there is a short circuit on the dead side. Another difficulty is that by holding the handle in an intermediate position the circuit can continue independently of the trip mechanism.

One way of getting over the trouble is to allow for a great deal of slack motion, so that when the trip levers are approaching their final position they can still be tripped. A better method is to bring the final lever over a fixed fulcrum point of the gear, and shape the mechanism in an arc round this point. By arranging a tappet or lever to trip on the curved portion, the gear will operate without slack in any position of the switch.

With automatic switches the motion of the moving element must be arrested without shock. When the moving parts are light a simple compression spring under a collar on the operating plunger does all that is required, but oil or air dashpots are employed in some cases.

A very simple arrangement for larger switches having the insulator bridge under the blade is to fix a plate to the moving element which comes up against another plate fixed at the bottom of the guides. Only a small mechanical clearance is left when the switch is " out," and the action of expelling the oil retards the motion smoothly (Fig. 5).

Strong springs are required to pull the blades out of the contact, but only a slight force in addition to gravity gives the necessary power for movement when free. The shock to be taken up by the buffer can be reduced without detriment to the working of the switch by arranging two springs in series with a lever at the junction. One spring is much stronger than the other, and the weaker spring does the work until the blades enter the contact. The lever between the springs is then brought up against an adjustable stop, so that the strong spring is brought into action for withdrawing the blades from the fingers. The more powerful spring thus overcomes the contact friction and gives the acceleration in the first instance. As the weaker spring is in action during most of the motion, the shock is much less than if only one spring were employed (Fig. 11).

Switch Position Indicating Gear. Some means of indicating whether the switch is in the " on " or " off " position is necessary, more especially when free handle mechanism is employed. A very simple method of doing this is to fit a contact on the switch to light red and green lamps for the two positions. The lamps may be supplied from a potential

transformer, but more usually a separate
circuit is employed. Sometimes only a red
lamp is used, but in this case if the filament
breaks or other trouble occurs, the switch is
shown " out " when it is really " in." Positive
indication of both positions is preferable.

Mechanical indicators are easily arranged
when the switch is behind the control board.
A very simple form consists of a light rocking
frame pivoted on some part of the switch
or in an indicator casting, and operated by an
adjustable rod from the switch mechanism
(Fig. 1).

Mechanical Remote Control. It is not always
convenient to place the switch gear close
to the operating panels, even when small sizes
are in question. When extra high tension
boards are required, the switch panels take up
a great deal of space, while it is an obvious
advantage for the control gear to be kept
small.

Remote control gear may be either mechanical or electrical. The mechanical gear consists of rods and bell cranks somewhat similar
to railway point rodding. The conditions
of operating are nothing like so onerous as

those experienced on a railway. The lengths and forces involved are not so great, nor is the temperature variation so extreme, and the gear is not exposed to the weather. The amount of red in tension and compression is equalized to avoid expansion troubles, when the length involved is over about 20 ft. (Fig. 7).

It is advisable to allow for a considerable range of adjustment, and in order to take up slack in the joints it is better to arrange for a little more motion than is actually required to operate the switch. The mechanism in the switch cell can be brought up against a stop at one end of its travel, the excess motion of the rod being absorbed by a strong spring. Adjustment at one end of the travel only is then required.

The release gear is sometimes fitted at the operating end of the rods, but in this case the tripping mechanism has to move the whole weight of the rod gear, and sluggish motion results.

If the trips are arranged in the switch cell, the trip wiring is lengthened, and an electrical indicator is required instead of a mechanical arrangement, but these are comparatively

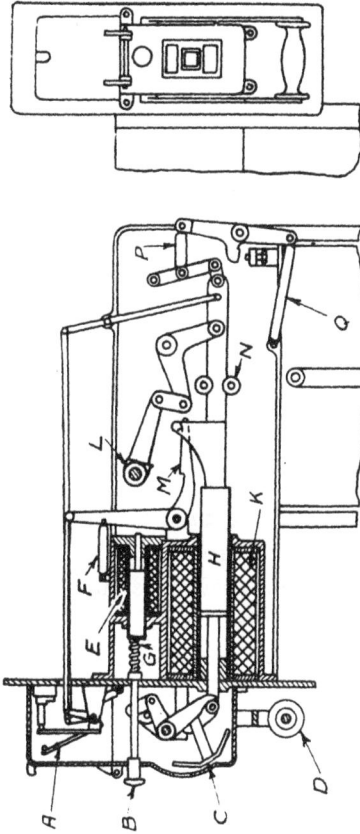

FIG. 11.—ELECTRICAL OPERATING GEAR FOR OIL SWITCH.

A. Switches for signal lamp and coils.
B. Hand trip.
C. Position indicator.
D. A hand lever to put in switch for testing or in emergency.
E. Trip coil.
F. Spring.

G. Spring.
H. Armature.
K. Pulling in coil.
L. Switch shaft.
M. Catch lever.
N. Roller.
P. Main springs.
Q. Auxiliary main spring.

minor disadvantages as compared with the advantage of a quick-break.

Electrical Remote Control. Electrical remote control is very largely in use, more especially for extra high tension work. The control board need be made only large enough to carry the instruments, and it may be arranged in any convenient position.

The most simple form of electrical gear consists of two solenoids. A large coil when excited closes the switch, which is held in the " in " position by a catch. A smaller coil trips the mechanism by moving the detent. A change-over switch is operated at the same time as the main contact and closes the tripping circuit when the main switch is " in," and the closing circuit when the main switch is in the " open " position, thus preventing the coils being burnt out by the exciting circuit being left closed (Figs. 5 and 11).

This switch also forms a convenient position indicator, lamps being placed permanently in series with the two coils : the " in " lamp in series with the trip coil, and the " out " lamp in series with the making solenoid. The resistance of the coils is not sufficient to reduce the brightness of the lamps to any extent.

The coils need not be designed for continuous rating. If the switch can be operated, say, five times consecutively without the temperature rising more than 90° Fahr. no trouble is likely to occur.

The switch closing coil should do its work without difficulty if the pressure rises or falls say 10 per cent., but it is an advantage if the tripping coil will operate even if the pressure falls say 25 per cent., as it is of great importance for the circuit to be opened when required of a severe fault. The values mentioned above for rating and pressure variation are those specified by the V.D.E.

The current taken by a solenoid gear is rather large, the momentary input of energy being as much as 3 kw. for quite a moderate size of switch. If a suitable source of power is available this is not a matter of great moment, as the energy is not required for more than a second, and so the watt-hour consumption is low.

In order to reduce the current taken, two solenoids are arranged in tandem in some designs, the second coming into operation after the first has done its work. A greater length of travel is given, and the actual pull required is reduced.

In one design of gear the closing of the switch is performed by springs, which are stretched by the action of an electric motor working through reduction gear. When the preliminary operation has been carried out, the circuit can be completed at any convenient moment by exciting a trip coil which releases the moving element. The switch is opened by another trip coil, the movement being by gravity, usually assisted by springs. This arrangement takes only a small current, but the switch cannot be closed a second time without an interval elapsing.

Current reducing arrangements do not actually reduce the watt-hour consumption of the gear, in fact the reverse is usually the case, as some loss of efficiency is generally involved. The only advantage is a reduction of the momentary peak, which is not of great importance unless the source of power is of very small capacity.

Solenoid gears suffer from the disadvantage that if the operator keeps the " make " contact closed, the circuit is completed again immediately the switch is tripped by the automatic features. This trouble can be got over by arranging that the " make " circuit is broken immediately the switch is closed, and cannot be remade without some other movement than merely keeping pressure on the " make " push-button.

In one design this is done mechanically. A gear is wound up by the switchboard attendant in order to make circuit, and this is tripped mechanically by operating the push-button after making contact for sufficient time to close the main switch. Another arrangement employs a series coil which breaks the mechanical connection between the push-button and the relay switch, so that the push has to be brought back into the " off " position again to reset the gear.

One of the defects of early designs of electrical operating gear was the time lag which occurred between the closing of the control switch on the board and the actual closing of the main circuit. This gave trouble when synchronizing generators, but the difficulty has been got over by reducing the inductance of the operating coils and lightening the switch-moving parts.

Base Connected Switches. Another form of oil switch consists of thoroughfare insulators brought through the bottom of a cast-iron tank, and carrying contact blades which engage with fingers carried on a moving element. The fingers may be connected together through series trip coils, or short circuited by a copper connection (Figs. 3 and 4).

The moving element moves upwards to break circuit, coiled springs giving the required motion in the case of an automatic switch. This type of switch gives a very good current-rupturing capacity for a given length of break, probably because the circuit is broken near the bottom of the tank under a great head of oil.

In this design the switch is sometimes made non-automatic, but with the finger blocks short circuited through fuses (Fig. 4). The fuse wire is threaded backwards and forwards through presspahn sheets, and presspahn dividers are placed between the phases. This arrangement gives some of the advantages of an automatic switch at a very low cost, and is much used for mining work, as all the sparking takes place under oil. Low melting point fuse wire is employed to avoid the chance of igniting the oil, the rating of the fuse being much increased over that in air on account of the better cooling effect.

Both in the automatic and switch fuse forms these switches have the advantage that isolating links can be dispensed with in many cases. The whole of the mechanism is carried in the upper part of the apparatus, and can be removed for examination, leaving the live contacts under oil.

An objection frequently raised to this type of switch is that it depends on the action of springs working against the force of gravity. This disadvantage is more apparent than real, as except when V or crushed brush contacts are employed few switches will act under the action of gravity alone, and in practically all cases the motion tends to be sluggish unless assistant springs are used. Correctly designed coil springs seldom give trouble, and the mechanism is invariably operated by two or more springs, of such strength that, even if one fails, the other will work the switch.

Early designs of base-connected switches gave trouble through leakage of oil due to defective cementing of insulators in the base of the tank. Sulphur and glass powder was much in use, this medium being poured hot. The contraction on cooling allowed oil to leak through, but the employment of litharge cement eliminates this trouble.

When used in mines a current transformer is frequently mounted in the oil tank of this type of switch, the casting being extended to give the necessary extra space. The only instrument needed for mining work is usually an ammeter, so that if the meter is mounted on the top of the switch the unit is entirely self-contained (Fig. 3).

Selector Switches. Change-over oil switches are also used. The most simple form has three thoroughfare insulators per phase. A blade is hinged below the centre insulator, and is in two portions at an angle of about 135 degrees, so that, when one side is in the closed position, the other is open. The blades are operated from cranks on a shaft rotated by a handwheel. One portion of the link between crank and blade consists of an insulator. This type of switch only gives one break per pole (Fig. 12).

Another form of change-over switch has two entirely independent moving elements, similar

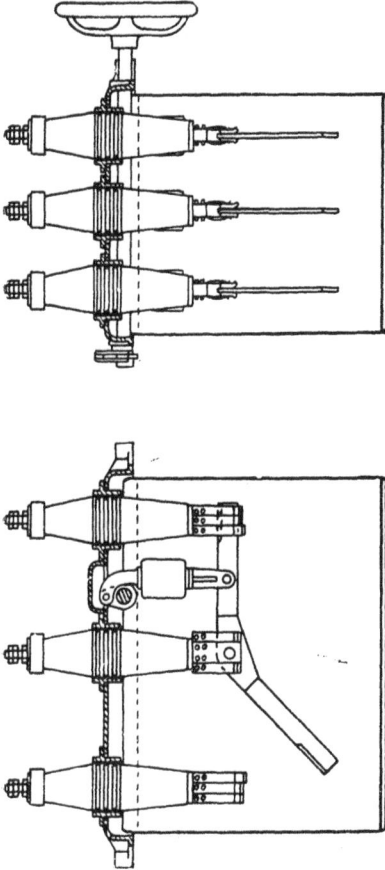

FIG. 12.—OIL-BREAK, CHANGE-OVER SWITCH.

to those in the conventional switch, but operated by one handwheel, so that when one is closed the other is open. This switch has a larger rupturing capacity than the single-blade form, as there are two breaks per pole. The centre contacts can be mounted on one thoroughfare insulator or four porcelains per pole can be used, allowing for two entirely independent switching operations. Both the single and the double blade types can be made with quick make and break, but they are more usually employed as oil break isolating links, with a main oil circuit breaker in series to protect them.

Another form of switch which can be used both for ordinary purposes or as a change-over switch has U shaped contacts, mounted on insulators on a rotating shaft. The legs of the U blades pass into clips mounted on thoroughfare insulators to make contact. This type is not now generally employed.

Oil switches can be applied for many special purposes : one design for furnace control has several blades similar to that on the single blade change-over design mentioned above. The operating cranks are connected to the main shaft by selected sliding keys, only one being in action at a time. The shaft has a hand lever working in a gate on the same principle as that used for motor-car speeds, and closes, one blade at a time, in either of two positions. By this means any number of tappings can be selected without the danger of closing two or more at once. The blades are held in the " off " position by some form of locking gear, as in an ordinary gearbox. This type is much superior to any form employing a brush sliding over a number of contacts, as current can be broken as in any other oil switch, while brushes tend to burn when called upon to open circuit.

Roller Mounting. Large oil switches are frequently mounted upon rollers so that they can be run out of their cells for examination after the connections have been removed. This arrangement does away with the necessity of leaving sufficient space round the switch to permit of examination, or minor repairs to be carried out ; and is especially convenient for the single-pole units in cell work where proper allowance for accessibility *in situ*

would often more than double the width of
the cells.   The single-pole units are frequently
operated by a rotating shaft having dismountable couplings so that the units can be handled
separately (Fig. 5).

CHAPTER III

TRIPPING MECHANISMS

Overload Trips. The most simple form of switch-tripping gear consists of a solenoid, usually mounted on the front of the board, and excited by a current transformer or auxiliary tripping circuit. The height of the plunger in the solenoid is made adjustable, so that the mechanism operates at various loads, except when an auxiliary circuit is in question (Fig. 1).

In order to get a time limit the coils are frequently short circuited by fuses. The solenoids are set to operate at a low current, the actual value at which the switch trips depending on the rating of the fuses. When fuses are employed the trips need not be calibrated, but it is an advantage to calibrate them in case it is desired to work without time limit in any instance.

The power available from this form of trip is not very great, depending as it usually does upon the output of a current transformer.

When fuses are not employed the readings of the switchboard instruments are affected injuriously, unless the impedance of the trip coil is kept low. This is especially the case with watt meters, as the power factor is varied by the induction effect.

The use of time limit fuses obviates the tendency to error, as the coil is only inserted when the switch is tripping after the fuse has blown. The energy consumption of the coil can be made greater than the output which the current transformer can give for any length of time without overheating, if this is desirable ; but in this case the trips cannot be used without fuses. With time limits the winding can be made very compact, as the tripping current is only carried for a few seconds at a time.

A typical design has 1,800 ampere-turns at minimum trip setting, operating on a laminated plunger approximately $\frac{5}{8}$ in. diameter. The coil is enclosed in a cast iron pot to complete the magnetic circuit, the reluctance of the plunger limiting the flux produced. This design is for use with current transformers having an output of 35 volt-amperes. The ampere-turns can be doubled with advantage, if fuses are used, by halving the section of the wire to get double the number of turns in the same space.

Directly wound solenoid trips are also used, but very little in the case of high tension, as the insulation of the winding has to stand the full bus bar pressure.

Other forms of time limits can be employed, such as an oil or air dashpot. In its simple form a dashpot time limit is only adjustable by varying the viscosity of the oil, but several patented variable time limits are on the market. In one form the plunger ends in a cone which fits into a similar cone at the base of the oil container. To obtain a variation in the time of tripping the plunger cone is left more or less out of the cone in the base, thus allowing oil to leak at varying rates to release the moving portion. Another form has a slot in the plunger, the size of which is varied by a portion of its length being blanked off by a part of the oil container. All dashpots give an inverse effect, or in other words, trip more rapidly with increasing loads, but the time limit varies with the viscosity and so with the temperature of the oil.

No-Volt Trip Gear. A very simple form of

no-volt attachment can be adapted for use with the same mechanism as the trip coils. A plunger is arranged to trip the switch under the influence of a spring, but is pulled out of operation by a pressure-wound coil. The coil has to be connected to the live side of the switch unless some arrangement is made to hold the plunger while the switch is being closed. One disadvantage is that since the coil acts directly on the mechanism, little energy is available for tripping unless the current consumption of the coil is excessive.

A more expensive form of no-volt mechanism, but one which offers considerable advantages, has the winding wound round a core which has a hinged armature forming one part of the magnetic circuit (Fig. 13). When the switch is closed a weighted lever is raised, and engages with a hook on the armature. The last motion of the switch releases the lever which trips the mechanism unless there is pressure across the low volt winding. This arrangement gives very much more power than the simple solenoid, but suffers from the disadvantage that the armature cannot lift the weight of the tripping arm, and so the magnetic circuit is held open if the coil

is excited when the oil switch is not closed.
When outgoing feeders are in question this
causes no inconvenience, as the coil can be
connected on the outgoing side ; but the case
of generators or ring mains necessitates the

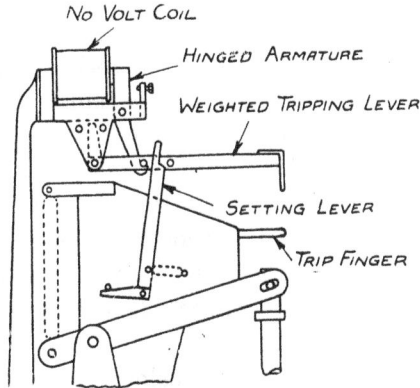

FIG. 13.—No-VOLT RELEASE MECHANISM.

winding being designed with a considerable
amount of ohmic resistance to prevent it
from being burnt out.

A late design of low voltage trip is arranged
so that the magnetic circuit is only open
momentarily when actually tripping the switch
(Fig. 14). The coil and armature are somewhat similar to those last described, but much

more iron and less copper are used. Great care is taken to ensure that there is a good magnetic contact over the whole of the pole faces, and a silencer winding is added.

The silencer winding consists of some eight turns of 18 gauge D.C.C. wire wound in slots on one of the pole faces, and encircling about one-third of the surface. In series with this winding about four turns of similar wire are wound in the same direction over the main exciting coil. Transformer effect causes a local circulating flux on the pole, when the main flux is passing though its zero value. As there is always a magnetic effect in existence the armature is never released and so chattering is avoided.

When the voltage falls below a certain value the armature drops under the influence of a spring and gravity, and releases a catch which holds a lever against the tension of a spring. This lever, when released, trips the switch and, in the same operation, pushes the armature back so that the magnetic circuit is again closed. The operation of opening the switch resets the lever, the armature being held in the closed position by a detent ; and when the switch is again closed the detent is pushed out of the way, allowing the trip to act unless there is pressure across the coil.

This arrangement has many advantages.

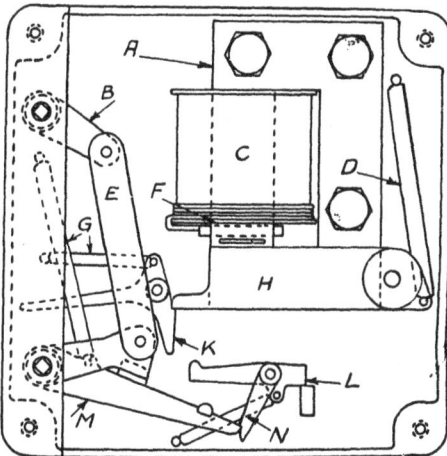

FIG. 14.—CLOSED CIRCUIT NO-VOLT RELEASE.

A.	Magnet core.	*G*	Springs.
B.	Setting lever.	*H.*	Armature.
C.	No-volt coil.	*K.*	Armature support catch.
D.	Spring.	*L.*	Trip lever catch.
E.	Link.	*M.*	Main trip lever.
F.	Silencing winding.	*N.*	Spring.

As the magnetic circuit is always closed there is no danger of burning out, although little copper is used in the coil. The power available is ample and the power consumption below 10 watts. The silence obtained is a great advantage for sub-station work.

No-volt coils should fall off at about two-thirds the working pressure, and hold on when the switch is closed up to 25 per cent. drop. No coil, whether pressure or current-wound, which is continuously in circuit, should heat up more than 90° Fahr. above the surrounding atmosphere. Fuse-shorted trips, when not calibrated, should operate five times successively without the temperature rise exceeding the above figure.

Direct Overload Releases. High tension series overload relays are often employed. This type dispenses with the use of current transformers, and makes the apparatus self-contained. Where current transformers are necessary for the instruments on the panel, no saving is effected ; but if this is not the case the cost is much reduced. An argument raised against the type is that there is likely to be flashing between the turns on short circuit, but this is rather a point against poor design than a criticism of the principle.

For switchboard type switches a coil is wound round a former, and placed round a core of iron stampings. A laminated armature is provided, to which is connected an insulated arm which trips the switch. The air gap of the armature gives a first adjustment, which is not varied when the switch is in use. The overload calibration is obtained by varying the tension of an adjusting spring.

Small switches are tripped directly by a horizontal arm, operating on a lever projecting from the trip case. In the case of the larger sizes a vertical tube is used and does its work through intermediate bell cranks.

For horizontal arms paxolin or some other form of compressed paper is the material in most general use, as this gives sufficiently good insulation up to about 6,600 v., and has good mechanical properties. The vertical rods have not such an onerous mechanical duty to perform, but as the pressures are generally higher the insulation must be better. In this instance mica or glass tubes are employed.

Oil-dashpot time limits can easily be applied to these overload releases. Another form consists of a clockwork mechanism employing a rack which rotates a small fan through gearing. This gives an inverse time limit effect, but the time limit is not easily made adjustable.

Another type of time limit utilizes the principle of the induction motor. When the armature is released it is held up by gearing operated by what amounts to a small motor, the field of which is provided by the main flux of the coil. After a certain number of alternating current cycles have been passed through, the revolutions of the motor completely release the armature which trips the switch. The time limit is not inverse, but can be varied by altering the number of revolutions of the motor which are necessary to allow the trip to operate. By cutting out the gearing when a certain pull is exerted on the armature the switch can be arranged to operate instantaneously on short circuit. This mechanism is somewhat expensive, as it requires in its manufacture the same degree of accuracy of workmanship as is generally used for instruments.

Direct overload releases can be arranged to operate under oil. An early form embodied a coil forming part of the bridge piece in the type of switch which has its insulators coming upwards through the bottom of the tank. The coil had an iron core which attracted an armature fixed at the side of the switch tank, behind an insulating partition. This arrangement was troublesome to assemble on account of the difficulty of aligning parts on the moving element with other parts on the tank, unless the air gap employed was excessive, and the insulation of the moving element was adversely affected.

A later form can be applied both to the same type of switch and to the more conventional form with the thoroughfare insulators coming through the top (Figs. 3 and 8). The armature is arranged vertically with its end extended to hold the piston of an oil dash pot time element when this is required. A trip rod with a reel insulator forming part of its length operates vertically.

The air gap is adjustable by a screw in a projection of the armature support. This trip, working as it does under oil, gets over the electrical objections to the relays working in air, as the insulation is fully equal to that of a current transformer.

In the case of the type of switch with the insulators at the bottom of the oil tank, the trip gear forms part of the short circuiting bridge, but with switches having insulators at the top a third support is necessary, though a simple pot insulator only is required. With the latter type the gear suffers from the disadvantage that the coils must be arranged at one side of the switch, so that the blades are not in the centre of the casting. This offers difficulties to the designer on account of the unbalanced lay-out of the switch.

Direct overload releases under oil are much used for mining switches. In installations used underground the instrument equipment is not usually extensive, and it is very desirable that current transformers should not be required merely to serve overload releases. In some designs an ammeter can be mounted on insulators, and the case left at bus bar potential, thus enabling a direct reading instrument to be employed. The instrument is read through wired or protected glass in the hood of the switch unit.

Independent Overload Relays. Relays are frequently independent of the switch, the actual tripping operation being either by a trip coil or by breaking the circuit of a no-volt release. A trip coil may be excited from an independent source of supply, by a shunt current from a pressure transformer, or by being placed in series with a current transformer on the switch panel. In the latter case the coil is normally short circuited by the relay, tripping taking place by removing the short. Series trip coils of this description are only really satisfactory when used in conjunction with overloads, as a reverse relay or other protective gear may be required to operate when little current is flowing in the mains.

Switches are often tripped through intermediate relays, in order to relieve the instrument contacts of the duty of making or breaking more than a certain current.

The most simple form of overload relay consists of a solenoid the plunger of which, when raised, makes or breaks the trip circuit. Several of these units can be mounted together to operate on a common trip bar to give double-, triple-, or four-pole effect. The most effective form of magnet is the pot form, but this gives rise to considerable inductance, and so when time limit fuses are not employed a form embodying a considerable air gap is used. This latter type lacks power, but makes a much reduced demand on the current transformer.

When time limit fuses are not employed, oil dashpots or mechanical time elements can be used, as with direct acting overloads.

The type of contact employed with these relays is usually of the non-arcing metal description. When a no-volt circuit has to be broken carbon contacts give excellent results, as the inductive arc causes no appreciable damage, and the correspondingly poor current-carrying capacity is of no moment as the ampere values are very small.

Other types of relays are designed on instrument lines. These have the advantage of great accuracy with the corresponding disadvantage of expense. Of these a very common form is the induction type of time limit overload relay. An aluminium disc operated on the Ferraris principle, as used for induction type ammeters, is arranged with its axis horizontal. In one arrangement there is a pulley mounted on the disc shaft, and on this there is a cord attached to a weight. The current at which the relay will act is adjusted by varying the weight, and the time limit is set by altering the distance through which the weight has to travel before the relay makes or breaks circuit. The relay is self-resetting by the action of the weight in revolving the disc backwards, when the current falls in the exciting coils.

Another type of induction overload relay has the exciting coils shunted by a resistance, current calibration being obtained by varying the amount of shunt. In this instrument also the time limit is adjusted by altering the motion required to operate. The discs in both cases are damped by permanent magnets, as in other Ferraris induction instruments.

For shunt-trips metallic contacts are employed, while for series-tripping mercury cups bridged by a copper link are sometimes used. Mercury cups give excellent contact with little friction, but the breaking capacity is small. Slightly more exacting conditions can be met by pouring a little oil on the surface of the mercury. Even with metallic contacts the breaking capacity of instrument relays is not very great, intermediate relays being necessary in many cases.

Induction type relays give an inverse time limit effect ; the greater the current, the greater the torque on the disc.

Almost any form of instrument can be used as a relay, the movement being employed to operate the contacts. Other forms of independent relays are adaptations of the ordinary series trips in which the mechanism is mounted separately from the oil switch and operates electrically instead of mechanically. The direct trip arrangement embodying the principle of the induction motor is especially successful where a time limit independent of the value of the current is required.

The time limit mechanism may be entirely independent of the relay. In one form the relay is arranged with two contacts. A normal overload current stresses a spring to a certain extent, and closes a contact which excites the operating coil of a time limit relay. After a certain time has elapsed, the time limit closes the trip coil, or opens the no-volt circuit of the switch. If the overload ceases before the time limit acts, the relay simply resets itself, without affecting the oil switch. The time limit is entirely independent of the actual value of the tripping current, as the time limit coil is excited from another source of energy. If a short circuit occurs, the control spring of the overload relay is fully stressed, and the second contact makes or breaks another circuit which trips the oil

switch immediately. The time limit used
with this arrangement may be either of the
clockwork and fan type, or the relay may
release a clock mechanism giving an absolutely
accurate time period.

Reverse Relays. Reverse relays are also
much in use. These may be of either the
reverse current or the reverse power types.
The reverse current form depends for its action
on the relation between two coils, one a current
coil and the other excited by the pressure.
When the current flows in the normal direction
these coils oppose one another, but when
power is reversed attraction occurs, and the
relay operates.

Reverse power relays are on the same
principle as watt-meters, and can be made
in single, two, or three phase form, balanced
or unbalanced. When the power is in the
normal direction the mechanism is brought
up against a stop, but on reversal the movement is free and opens or closes an auxiliary
circuit which trips the switch.

A difficulty with reverse relays arises from
the fact that if a serious fault occurs on a
generator, the voltage falls to a very low

value and so, although the reverse current may be very considerable, there is very little power to operate the relay. Instrument makers generally specify to what extent the pressure may drop and still allow the relay to perform its function.

When reverse relays are used, three phase generators are sometimes protected by one single phase relay only, or a three phase or three single phase instruments may be employed. A single phase relay protects against all mechanical faults, and against any defect in the exciter circuit, or rotor system generally, but does not act on a stator breakdown between the two phases not in circuit. It is an open question whether the extra cost of protecting the other phases is necessary to guard against this one contingency.

Special Relays. Minimum current trips are also employed in some instances, notably in connection with pumping plants, where they are arranged to open the circuit when the water supply ceases. These may either be direct wound or of instrument type. In the direct wound form an armature is held up by a detent, being placed in position by the action of closing the switch. The armature is not quite up to its pole piece until it is raised by the current flowing. When this occurs the detent moves away, allowing the armature to fall, and trip the switch when the current again drops. A mechanical arrangement may be provided to hold up the armature while the water valves are opened if the pump is started up under light load conditions.

Other types of relays are used in conjunction with special protecting systems. These often take the form of very sensitive intermediate relays making or breaking circuit when a small auxiliary current is made or broken. Interlocked reverse relays are also employed for feeder protection. The purpose of these types more properly comes under a consideration of switchboards.

CHAPTER IV

ISOLATING LINKS

Single Pole Links. The most simple form of isolating link consists of a single-pole, front-connected knife switch mounted on insulators (Fig. 15). On account of the pressures involved the break has to be much more than on the ordinary low tension switch for mounting on slate, and so the difficulty of aligning the blade and outer contacts is greater. For this reason clip contacts are more in use than the strip type almost universal for low tension work, as the blade is better guided into the clip. The smallest size made by most manufacturers is for about 100 amp. as otherwise the blade becomes too weak for its length. For very high pressures blades are sometimes reinforced by steel brackets so as to give a wider bearing at the hinge and a stiffer blade.

Single-pole isolating links are occasionally fitted with insulated handles, but they are generally arranged for pole operation. A metal hook mounted on an insulator fixed to the end of a wooden pole provides a convenient means of moving the blade.

The hook may work in a slot cut in the end of the knife, or in a hole either at the end of the blade or between the contacts. A much better method is to rivet a pressed mild steel

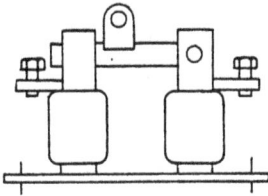

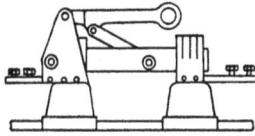

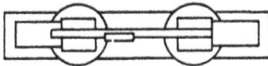

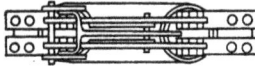

FIG. 15. FIG. 16.
SINGLE-POLE ISOLATING LINKS.

loop to the blade, as this gives a large hole for the hook to enter without reducing the current-carrying capacity of the switch.

It is not possible to get the same power to operate the link from a hook as can conveniently be obtained from the handle of a low tension knife switch. For this reason large isolating links are sometimes fitted with an auxiliary lever for opening and closing, a link motion being provided which approaches dead centre as the blade enters the clips (Fig. 16). As with other forms of knife switch, several blades are used in parallel for the larger currents, but brush contacts are used by some manufacturers.

As in low tension practice, current is sometimes conveyed through the hinge, but in other cases a separate clip is employed. Single- or double-throw assemblies can be made.

The contact clips are generally mounted on brass blocks, the current being brought out at the ends in the single-throw type or from the side in the case of the centre contact of a double-throw link.

Small links are mounted on plain pot insulators, but for high pressures the size becomes too great for a solid insulator, and hollow forms are employed which are cemented into cup-shaped cast-iron flanges at the base and top. As for oil switches, plain porcelains have replaced the corrugated forms. Thoroughfare insulators can be used when it is convenient to bring the connection from the back of the link.

Isolating links are frequently fitted with catches to hold them in the closed position.

If the link forms part of a loop in the conductors there is a very considerable force
tending to open the blade under short circuit
conditions, but if the current path is comparatively straight, the opening force is much
smaller.

It is sometimes necessary to employ isolators
out of doors in overhead line installations.
Outdoor links are mounted on special shed
insulators so designed that, under any weather
conditions, some part of the porcelain remains
dry and so retains its insulating properties.

Multiple Links. Double-, triple-, or four-
pole links are often employed, and offer considerable advantages when interlocking features
are required on a board. One form has the
blades fixed at their centre to a rotating shaft
of which portions are formed by porcelain
insulators. When making circuit the blades
fit into contacts on separate insulators, and a
double break is obtained in the off position.
The shaft is rotated by a handwheel at the
front of the board.

Another type has ordinary hinged blades
with links attached at about the centre of the
blade. An insulator forms part of the link,

the whole being operated by levers from the switch panel (Fig. 17).

Another form has the blades mounted on a moving element somewhat similar to that of an oil switch, but mounted horizontally instead of vertically. This type gives a double break per pole, but is somewhat expensive to construct (Fig. 18).

The blades of a multi-pole link are sometimes mounted on a base which can be taken right away from the switchboard, so that the circuit cannot be made when apparatus is under repair. Triple-pole link handles are frequently arranged so that they can be padlocked in the " off " position to secure the safety of the repair staff.

Commutator Links. Rotary type links can be adapted very easily to give two or more " ways," and the blades in the hinged form can be extended behind the hinge to perform another switching operation, but the switch in the latter case becomes somewhat bulky.

The straight-line motion form of links gives a very compact design for switching several circuits. The blades can be arranged to enter a set of contacts at either end of their motion

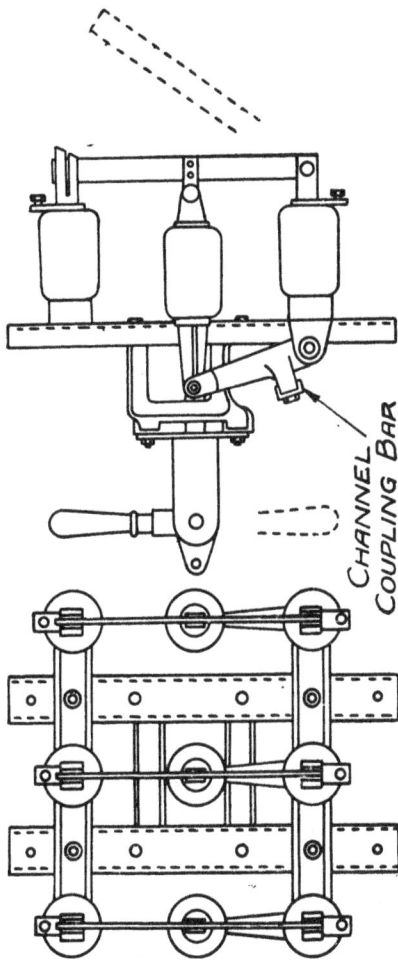

CHANNEL COUPLING BAR

Fig. 17.—Triple-pole Isolating Link; Design to Obviate Cross Straining of Insulators.

or, by stopping the moving element in intermediate positions, other connections can be made. One advantage is that each switching operation is entirely independent, which is convenient for connecting tappings on a three phase to two phase Scott-connected furnace transformer, and in other cases (Fig 18).

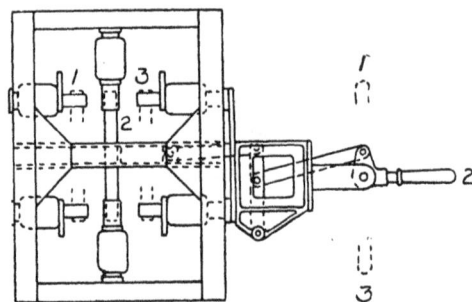

FIG. 18.—THREE-WAY TRIPLE-POLE ISOLATING LINK.

Sliding brushes are sometimes employed, making circuit between studs mounted on insulators and an insulated sector. This form is very occasionally used for switching potential transformers or electrostatic instruments on to one of several high tension sources of supply. This type offers difficulties of construction on account of the space between the contacts, and it is troublesome to use because of the difficulty experienced in finding the position of the relatively small contacts.

Apart from change-over switches proper, isolating links are frequently fitted with earthing contacts. Sometimes the link has three positions, " on," " off," and " earth." Earthing clips can be added very easily to almost any form of isolating links, as no additional insulators are required. Although the contacts must make absolutely certain earth connection, they are not required to carry current continuously, but the same form of clip as that used in the main switch is generally used.

Remote Control. It is sometimes convenient to operate isolating links by rod gear, so that they may be placed in inaccessible positions, but electrical gear is never used, as rapidity of operation is of no importance. The essential purpose of isolators is to enable other gear to be examined or repaired without risk to the mechanics, and so simplicity and robustness of construction are very necessary.

Mast Switches. Knife switches were formerly used to break high tension circuits, and the earliest type of oil switch had only the sparking gear under oil, the main current being carried by switch blades in air. The only survival of the air break high tension switch is the mast switch used for overhead line gear (Fig. 19).

In a mast switch two sparking horns are provided per phase on two fixed insulators, of which one is connected to the incoming side and the other to the outgoing side. A third, moving insulator is provided, and this has a contact clip attached to it and connected to one of the fixed insulator caps by a flexible connection. The other fixed insulator has a blade over which the clip fits to make circuit. When breaking current the clip of the moving insulator passes along the base of one horn, and then along that of the other, and so transfers the sparking to the horn gap. Under the influence of the magnetic forces an arc across a horn gap is not stable, but tends to rise to where the horns are wide apart, and so break circuit as the sparking distance becomes too great for the arc to be maintained. At one time this type of switch was much used in central stations, but it has been discarded in favour of the oil switch. For use on overhead

lines the mast switch has the advantage of
low cost and, unlike the oil switch, it need not
be enclosed. It requires little attention, and
the large flash which occurs on breaking circuit

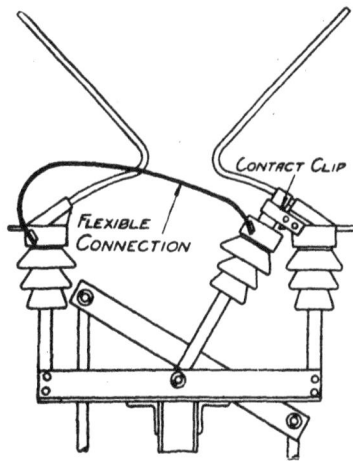

FIG. 19.—MAST SWITCH.

can do no damage out of doors. The type still
survives for breaking sub-circuits in the open
air, but not for the main switching operations
in the generating and sub-stations.

CHAPTER V

General Considerations. Arrester gear is designed to protect an installation from the effect of lightning, from over-pressure caused by inductance and resonance effects, or to prevent pressures from the high tension side of a transformer reaching the low tension side.

A lightning discharge is of the same nature as the discharge from a condenser in the familiar influence machine experiments. The voltage is very great, the frequency very high, and there is very considerable power to be dealt with.

The fact that the frequency is so high is made the basis of all protective gear. The lightning discharge will not pass through a choking coil of even very small inductance if an alternative non-inductive path is provided, and thus the discharge can be removed before it reaches any apparatus which may be damaged.

Choking Coils. The most simple form of

choking coil consists of a few turns made in
the connection to the overhead line itself,
but separate choking coils are usual. For
indoor use with small currents and for pressures
up to 6,600 v., some 10 or 12 turns of copper
of not less than $\frac{3}{8}$ in. diameter are wound to
form a coil about 4 in. diameter with about
1 in. between turns (Fig. 20). The coil is
mounted upon insulators and one end is

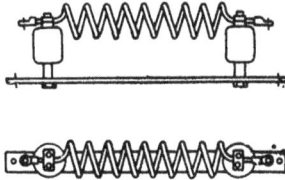

FIG. 20.—INDOOR CHOKING COIL.

connected to the overhead line and the other
to the gear to be protected, the main current
passing through the coil. For outdoor use
shed insulators are employed and the winding
diameter and spacing of turns are increased
to about 6 in. and 1½ in. respectively for the
working conditions mentioned above. In this
case the coil is generally arranged with its
axis vertical, while indoor types are mounted
in any position required for convenience.

For larger currents and pressures the dimensions are increased. One design for 10,000 v. and 600 amp. has 24 turns divided into two coils of 12 turns each, the whole being mounted on three insulators. To save the expense of copper, iron is used, the current density being kept very low. Iron introduces considerable magnetic effect and, to prevent this from giving trouble, insulating blocks are placed between turns and the whole is bound together to give rigidity.

Choking coils are also used in bus bar layouts and in generator connections to reduce the rush of current on short circuit. This type is much more rigidly constructed than those used for lightning protection, marble and other materials being used for mounting.

Horn Arresters. The lightning is prevented from reaching the board by choking coils only if some form of discharge gear is employed to provide an alternative path. For high tension circuits the most simple form in general use consists of two horns of copper or galvanized iron, one mounted on an insulator and the other earthed. The gap between the horns is arranged to give a factor of safety for the

ordinary pressure of the line, but breaks down
when the lightning strikes. An arc between
horns is unstable, and so the current to earth
ceases when the high tension energy of the
lightning has been discharged. The horn

FIG. 21.—OUTDOOR HORN ARRESTER.

gaps are of course connected on the line side
of the choking coils, the connections being
kept as straight as possible, so that the high
frequency current is not opposed by any
inductance.

The simple horn arrester has the defect that
when a discharge takes place the line for the
time being is earthed, with corresponding
shock to the generating plant. To obviate

this trouble both horns of the arrester are mounted on insulators, one side being connected to earth through a non-inductive resistance of about 1 ohm per volt of normal line pressure (Fig. 21).

Non-Inductive Resistances. The non-inductive resistance may take the form of slabs of resistance material composed largely of carborundum. Each slab is approximately 6 in. by 3 in. by ½ in. thick, and has about 1,000 ohms resistance. A supporting insulator is provided between each two slabs, and the whole is mounted on an iron strap. This type is very convenient for mounting in a switch board.

Another form employs asbestos and wire resistance-strip similar to that used for oil switch charging resistances. The strip is placed in an oil tank, being wound on reel insulators alternately from the top to the bottom of the tank and vice versa. One end is connected through a thoroughfare insulator to the horn gap and the other end is earthed.

The type most used in England is the carbon dust resistance (Fig. 22). The resistance medium is a mixture of carbon dust and carborundum powder, the proportion being varied to give the required resistance. The dust is placed in grooves in earthenware troughs, the troughs being mounted one on

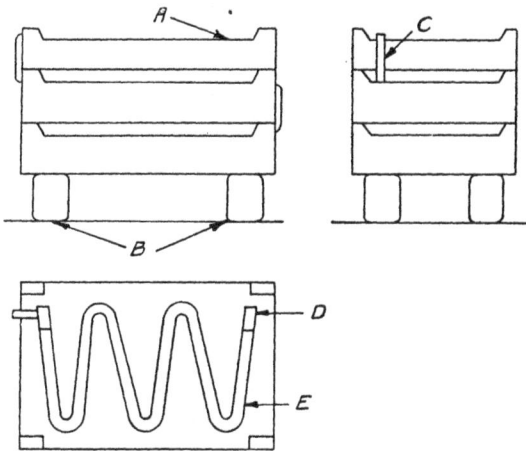

FIG. 22.—CARBON DUST RESISTANCE.

A. Earthenware trough. D. Carbon block contact.
B. Porcelain insulators. E. Carbon dust in groove.
C. Connection between troughs.

top of the other, and lugs being provided to give a clearance. The whole bank of troughs is mounted on insulators. Connection is made by carbon blocks with flexible copper connections embedded in them similar to dynamo brushes. This method is very successful, and has the advantage of a negative temperature co-efficient of resistance, so that the larger the current passing, the lower the ohmic resistance offered.

The same general type is also used for earthing the neutral point of generators, when this is considered desirable. For this purpose the carbon dust is generally the whole width of the trough, and not in zig-zag grooves, as the current to be carried is much greater than for lightning discharge. Troughs are mounted on insulators as for arrester gear, but they are frequently connected series-parallel to deal with the larger currents.

Application of Horn Arresters. Horn arresters are frequently connected with resistances in series between each phase and earth, but sometimes instead of directly earthing the common point, it is connected through another gap and resistance to the earth plate. The latter arrangement gives two gaps and resistances between each phase, and two gaps and resistances between each line and earth.

Horn arresters are frequently used out of doors on overhead line installations, the horns being mounted on shed insulators and increased in dimensions as compared with those for indoor use. For instance, two designs for use up to 10,000 v. and for indoor and outdoor service respectively each have horns of galvanized iron, $\frac{3}{8}$ in. in diameter and so mounted that the spark gap can be anything from zero to 1 in. The distance between the tips of the horns is approximately 12 in. for indoor and 21 in. for outdoor service, and the corresponding heights of the horns from the contact-blocks are about 6 in. and 14 in. respectively (Fig. 21). If used out of doors, carbon dust resistances must be protected, and cast-iron boxes are generally used for the purpose.

Excess Pressure. Protective gear is also used to prevent excess pressure on the system, more especially when the neutral is not earthed. When a large cable system is in operation, the condenser effect of the cables tends to keep the pressure within limits, but in all cases there are conditions under which the voltage rises considerably. Resonance effect and surges caused by switching or breakdowns are the chief causes of the trouble.

Horn arresters are used to guard against

excess pressure, but they are not very successful for the purpose. With lightning discharge the voltage is many times that of the line and the gaps can be set accordingly, as the more delicate gear is protected by the choking coils, but it is not desirable to allow the bus bar pressure to rise to this extent. If the gaps are set small the presence of dust or steam may cause discharge at normal voltage. To get over the difficulty, in one design, the main gap is set with a considerable factor of safety, but there is an auxiliary gap between the horn at bus bar voltage and a point connected to the horn at the lower potential through a non-inductive resistance. This type, however, is not in general use.

Gap Arresters. Gap arresters are employed extensively for over-pressure discharge. These consist of cones mounted on a marble base, the whole being supported by insulators (Fig. 23). The cones are arranged alternately pointed towards and away from the base, and supported by threaded stems, so that the distance from the marble can be adjusted. The stems are insulated from the marble by mica plates and bushes to guard against the

danger of metallic veins and faults in the
material. One cone is connected to the line
and another to earth, generally through a
resistance ; the remainder being arranged
in series with a small air gap between each.
By raising or lowering the cones on their
stems, the gaps can be adjusted. When excess
pressure occurs, the current jumps across the

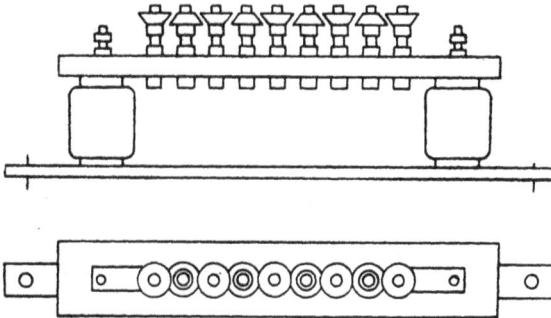

FIG. 23.—CONE ARRESTER.

gaps and discharges itself. Owing to there
being many small breaks, the heat developed
in each one is not sufficient to vaporize the
metal to any extent, the design giving a comparatively large mass of material to be heated.
As little vaporization occurs the gaps cease to
discharge as soon as the pressure drops, and

the cones are not burnt to any extent. Any slight defect on the surface of the metal caused by discharge can be made good by rotating the discs on their spindles.

A typical design for pressures up to 6,600 v. has 9 brass discs placed at $1\frac{1}{8}$ in. centres, the cones being $\frac{3}{8}$ in. thick and $1\frac{1}{8}$ in. at their larger diameter, the faces being at an angle of 45 degrees. For other pressures the number of cones is varied proportionally.

Since the line is earthed when discharge occurs, non-inductive resistances are generally used in series. As with horn arresters, either three or four sets are employed to protect a 3 phase installation.

Water Jet Arresters. Water jet arresters are also adopted to protect against surges. This type of arrester has the effect of earthing the line permanently through a high resistance, and so gives a continuous discharge. In one form inverted cones, one for each phase, are mounted on shed insulators. Each cone is arranged over an adjustable water jet, the liquid spraying on the inside and dripping off the edge. The whole installation is mounted on an earthed water tank, the cones being connected to the lines. As the liquid is changed continuously no overheating occurs, although the water column is small in dimensions.

One design on this principle for use on 10,000 v. lines, has cones $4\frac{1}{2}$ in. diameter and $3\frac{1}{2}$ in. deep, arranged at a height that can be adjusted

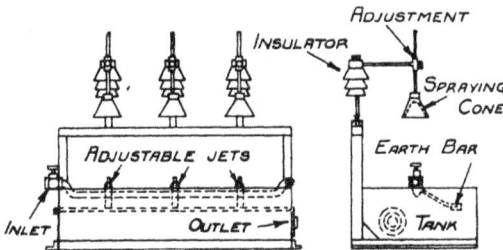

FIG. 24.—WATER JET ARRESTER.

up to 1 ft. above water jets which are about $\frac{1}{16}$ in. diameter at the orifice (Fig. 24).

Another form of water jet arrester employs two glass tubes of different sizes, mounted vertically one inside the other. A metal cap on the top of the outer tube is connected to the line, and water circulates up the inside and down the outside tube, thus giving the requisite change of water.

Other types are in use, notably one employing electrolytic action between a number of aluminium electrodes. The action offers a back pressure to the line, the resistance being broken down by any excess value. On discharge ceasing the bus bar volts are again opposed.

Break Down Arresters. Another form of protective gear is that placed on the low tension side of a transformer. All insulation must be designed with a considerable factor of safety, in fact many makers test their low tension gear to at least 2,000 v., and some use a minimum of 10,000 v. for all apparatus except instruments. If a transformer breaks down between windings the pressure may be transferred to the low tension side without the high tension breakers being affected, but the danger to the operators from high voltage charge on comparatively lightly insulated low tension apparatus is very considerable. Break down spark gaps are designed to guard against this danger. These consist essentially of a weak point in the insulation, so arranged as to break down when excess pressure is applied, but not to give trouble under normal conditions. When the gaps discharge the low tension system is earthed temporarily, and the high tension breakers then trip.

The most usual form of break down spark gaps consists of discs of zinc or other non-arcing metal, arranged one above the other alternately between mica sheets, so as to give a small spark gap. The gap is split into several small parts, and the bulk of the discs and nature of the metal ensures that the arc will not be maintained when excess pressure ceases (Fig. 25).

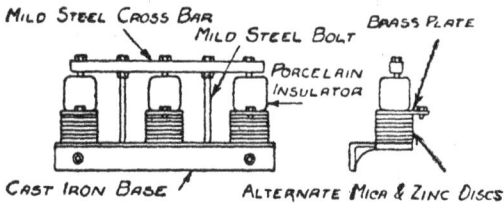

FIG. 25.—TRIPLE-POLE BREAK DOWN SPARK GAPS.

A typical design employs zinc discs about $2\frac{1}{2}$ in. diameter by $\frac{1}{8}$ in. thick, alternately with mica discs of the same diameter but $\frac{1}{32}$ in. thick, three micas being employed for 500 v. normal pressure. Current is conveyed to the apparatus by a brass plate with a lug for bolting the connection. The whole being mounted on an earthed base and clamped on an insulator, kept up to its work by a set screw.

8—(5331)

For mining work such spark gaps are frequently mounted in a cast-iron box with wide machined faces to render them flame proof. Lamp sockets are sometimes fitted both to the enclosed and to the open type, the connection being across the gap. A lamp can be inserted to test if the gear is in working order, the filament glowing when there is pressure across the apparatus.

An enclosed type is also used. The discs in this form are made with bevelled edges, so that the spark passes through the mica and not over the edge, and the whole is enclosed in an insulating tube. The ends of the tube are arranged to fit into fuse clips, which form a convenient mounting.

CHAPTER VI

FUSES other than oil switch fuses are not used to anything like the same extent for high tension work as on a low tension system, but they are employed in some cases where the expense of an oil switch is not justified. The energy set free in breaking a high tension circuit is so considerable that the design offers difficulties.

Potential transformers are always protected by fuses, generally supplied by the instrument maker. These usually consist of glass tubes containing a long length of very small fuse wire, sometimes packed in chalk dust or other material to keep down the arc. In other types the fuse is under the tension of a coil spring, so as to give a long and quick break. As the current to be broken is very small, such fuses are relatively easy to design despite the high pressures dealt with.

The most simple form of high tension fuse to carry an appreciable current consists of an

ordinary porcelain handle fuse of much larger
dimensions than the low tension type. The
porcelain is very substantial as the heat
developed in breaking circuit is considerable,
and the fuse wire is frequently enclosed in an
asbestos tube. Small horn breaks are sometimes provided on the top contact clips to
take the slight spark which occurs on breaking
circuit, owing to capacity effects, even when no
appreciable current is passing in the mains.
The contacts are protected by the wide flanges
to prevent accidental touching of live metal,
and as an additional precaution wooden tongs
are sometimes used to withdraw the fuse when
live. Such fuses can be used up to about
100 amp. on 6,600 v. circuits (Fig. 26).

Another form has the fuse wire in a carrier.
The fuse element is inserted in an inverted
pot-shaped insulator, suitable vents being
provided for the escape of gases generated on
breaking circuit. This type is designed for
use out of doors.

Oil is employed in some fuses other than the
oil switch fuses already considered. An early
form had two compartments with a fuse
brought across the partition between them.
The wire was attached to springs which drew

the ends of the fuse under the oil on breaking
circuit and so extinguished the arc. Glass
plates were placed above the troughs to prevent
trouble from dust. When these fuses opened
on short circuit very considerable sparking
occurred.

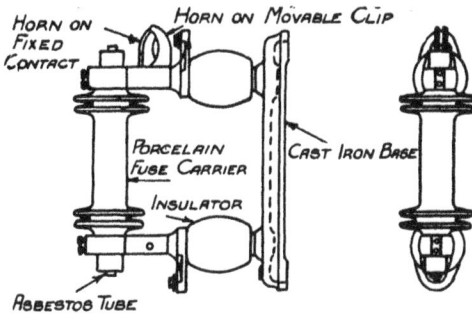

FIG. 26.—HIGH TENSION FUSE.

Another type consists of a glass tube with
a fuse wire attached to one end cap. The other
end of the metal is soldered to a coil spring
which draws away the molten fuse and gives
a long break.

Horn breaks are also employed. The fuse
is placed across the base of the gap, so that the
arc can run up the horn. In some cases the
fuse carrier is hinged and fitted with contacts

at either end, making circuit with blades on the gap. When the fuse has to be renewed it is lowered on the hinge and is then dead, independently of the condition of the main circuit.

CHAPTER VII

TESTING

Current Testing. Testing of switchgear divides itself into two portions, testing for current carrying capacity and testing for insulation.

For current testing a special transformer is generally employed. The primary is fed from a convenient source of supply, and the secondary is wound to give a large current at a low pressure. For high tension work an apparatus giving 2,000 amp. at about 8 v. is a convenient size.

If a special transformer is not available instrument current-transformers can be used. Several transformers are generally necessary, the primaries and secondaries both being connected in series, care being taken that the coils are in the same direction on both sides, *i.e.*, if the primary current flows from L to N, the secondary must also be arranged with flow from L to N. The nominal ratio of the current transformers does not hold good, and a point is soon reached when increasing the supply current has no effect on the output, but if good contact is secured on the heavy current side, quite good results can be obtained.

For current testing, the transformers are almost invariably connected through a rheostat on the supply side, the resistance on the secondary being kept as low as possible. This arrangement results in economy of energy as compared with any method of using a resistance in the path of the heavy current. The current is of course metered on the low tension side.

A much smaller range of resistance is required than would appear necessary at first sight, as apparatus for small currents has generally a much larger contact and conductor resistance than the larger type, so that the energy supply is not very different in the two cases.

Current testing quickly reveals the large factor of safety which is involved in using the accepted 1,000 amp. per sq. in. rating for small ampere values, but when using currents exceeding 1,000 amp. great care is necessary in making connections, if heating trouble is to be avoided. Despite the low pressure a shock can be obtained when the secondary

of a current testing transformer is opened
owing to inductive effects, but no danger is
to be feared.

As regards heating, variation in frequency
causes no difficulty, but if overload coils are
to be calibrated each switch, strictly speaking,
should be tested on its own frequency. However, the error involved in using a fixed
periodicity of say 50 cycles, is not very great,
it having been found from actual tests that
direct overloads calibrated at 50 cycles are
correct within 2 per cent. on continuous
current or alternating current of 90 cycles
per sec.

The V.D.E. allowance for overload calibration is plus or minus $7\frac{1}{2}$ per cent.

The calibration of transformer-operated relay
and of no-volt coils is fairly simple. If a
motor generator or inverted rotary converter
is not available, an allowance for frequency
has to be made when calibrating no-volt coils.
If the coil be assumed to be all inductance,
halving the cycles has the same effect as
doubling the volts, hence, if a 110 v., 25-cycle
coil will run without overheating on 220 v.,
50-cycle supply, and will hold on without
trouble at a value not greatly in excess of the

maximum pressure allowed for a 50-cycle coil, it can safely be passed ; but as the assumptions made are very approximate, a good margin should be allowed. Of course nothing in the way of instrument calibration can be done on these lines. The ohmic resistance of no-volt and other coils is generally measured to detect shorts between turns.

Pressure Testing. Insulation is tested by subjecting the apparatus to a pressure considerably in excess of that to which the insulation is intended to be stressed in practice.

Pressure tests are generally taken to earth and between phases, the voltage used being the same in both cases. No-volt coils and any pressure-wound instrument that may be used are usually disconnected when testing between phases, but not when testing to earth.

Some makers specify that all their gear is tested to one and a half times working pressure, with a minimum of 1,500 v. or 2,000 v. for apparatus on low tension circuits, but the V.D.E. values are much in excess of this, and are as shown in the Table on page 113.

TEST PRESSURES SPECIFIED BY V.D.E.

Working pressure	Short-circuit current	Test pressure
Volts	Amperes	Volts
1,500 3,000	Up to 6,000 } Up to 3,000 }	10,000
3,000 6,000	2,000 to 6,000 } Up to 2,000 }	20,000
6,000 12,000	2,000 to 6,000 } Up to 1,500 }	30,000
12,000 24,000	1,500 to 4,500 } Up to 1,000 }	50,000
24,000 35,000	1,000 to 2,000 } Up to 1,000 }	70,000
50,000 80,000	Up to 1,000 Up to 1,000	100,000 160,000
110,000 150,000 200,000	Up to 1,000 Up to 1,000 Up to 1,000	220,000 300,000 400,000

The V.D.E. specification adds that oil switches are to be tested to earth when open and closed, between phases when closed, and between parts of the same pole or phase when open.

It will be noticed that the size of the

installation, as well as the working volts, are specified.

Special test transformers giving a sufficient · output to break down an insulator should be employed for the final tests, although instrument potential transformers are sometimes convenient for a preliminary examination when a fault is expected during erection.

An electrostatic voltmeter is usually employed to read the test volts, but the conditions on the low tension side give some indication of the state of the insulation, as any breakdown causes the instrument needles to be unsteady, owing to the instability of an arc.

The pressure should be increased gradually to avoid sudden stress on the insulation, and the voltage should be maintained for at least one minute so that a fault has time to develop. Any convenient frequency may be employed independently of that for which the apparatus is designed. The V.D.E. specify 50 cycles as a standard.

The wave form given by a test transformer should be approximately a sine curve, as the insulation is tested by the maximum value of the pressure and not by the root mean square value as read by an instrument.

Before testing, it is usual to read the value of insulation resistance with a megger or other testing instrument; the figure obtained for individual pieces of apparatus designed for high tension work nearly always approaches infinity.

The connection between the test transformer and the gear to be tested is generally made by small fuse wire, as the soft metal gives a good connection, while if an earth occurs the wire burns away without damaging the transformer.

Considerable brush discharge occurs with the higher pressures, but this is not dangerous in most instances. If the pressure seems likely to jump across a gap, a flame, such as that of a lighted match, thrown between the points will often cause a flash over. If the arc persists the sparking distance is too small, but otherwise the factor of safety is sufficient when the difference between the test and working pressures is taken into consideration.

BIBLIOGRAPHY

FURTHER information on the apparatus and practice described in this volume and in the companion primer on *High Tension Switchboards* may be obtained from the following papers—

" Merz-Price Protective Gear and other Discriminative Apparatus for Alternating Current Circuits," by K. M. Faye-Hansen and G. Harlow. *I.E.E Journal*, May, 1911.

" Pressure Rises," by W. Duddell. *I.E.E. Journal*, 1st Dec., 1913.

" Horn Lightning Arresters," by C. C. Garrard. *The Electrician*, 20th Mar., 1914.

" Current Limiting Reactances on Large Power Systems," by K. M. Faye-Hansen and J. S. Peck. *I.E.E. Journal*, 15th Apr., 1914.

" Ferranti Waters Protective System." *Electrical Engineering*, 25th June, 1914.

" Split Conductor Mains Protection." *Electrical Engineering*. 22nd Oct., 1914.

" Over-Pressure Protective Gear for High Tension Circuits," by K. Edgcumbe. *Electrical Review*, 27th Nov., 4th, 11th, and 18th Dec., 1914.

" Automatic Protective Switchgear for Alternating Current Systems," by E. B. Wedmore. *I.E.E. Journal*, 15th Jan., 1915.

" Overload Protection on Alternating Current Circuits by Tripping Devices," by C. C. Garrard. *The Electrician*, 29th Jan., 12th Feb., 1915.

" Switchgear Standardization," by C. C. Garrard. *I.E.E. Journal*, 18th Apr., 1918.

" Control of Large Amounts of Power," by E. B. Wedmore. *I.E.E. Journal*, 18th May, 1918.

At the time of going to press the British Engineering Standards Association has in preparation a series of standard specifications covering various types of air-break and oil-immersed switches and circuit breakers.

INDEX